Workbook for

Video

Digital
Communication
& Production

Second Edition

By

Jim Stinson

Publisher

The Goodheart-Willcox Company, Inc.

Tinley Park, Illinois

www.g-w.com

Contents

Video Digital Communication & Production Workbook **3**

	Text	Workbook

About Video

Reading Review

Write a brief answer for each question below. If you are not sure of an answer, review the appropriate section of the chapter to find it.

1. Name the two media that are the "parents" of video.

   _____

2. Name some ways in which film is superior to video.

   _____

   _____

3. Name some of the drawbacks of film.

   _____

   _____

   _____

4. Explain how you can respond to video if you are "visually literate."

   _____

   _____

   _____

5. Explain how the video screen may be thought of as a "window."

   _____

   _____

6. Describe some ways in which the video world differs from the real world.

   _____

7. Name the four major elements of the language of video expression.

   _____

8. Relate each of the four elements to elements of spoken or written language.

   _____

   _____

9. Name the three major phases (parts) of video production.

   _____

10. Video editing is said to be very satisfying creatively. Why do you think this might be the case?

    _____

    _____

    _____

Vocabulary Review

Match each term to its definition by writing the letter of the correct definition in the space provided. (Not every one of these terms is defined in the *Technical Terms* section in the textbook, and some of the definitions do not apply to any term listed here.)

Terms

_____ 1. Broadcast

_____ 2. Camcorder

_____ 3. Digitize

_____ 4. Film

_____ 5. Gray scale

_____ 6. High definition video

_____ 7. Image

_____ 8. Live

_____ 9. Postproduction

_____ 10. Preproduction

_____ 11. Production

_____ 12. Resolution

_____ 13. Scene

_____ 14. Sequence

_____ 15. Shoot

_____ 16. Shot

_____ 17. Television

_____ 18. Video

_____ 19. Video world

_____ 20. Visual literacy

Definitions

A. Editing a program on computers and other digital equipment.

B. An audiovisual medium that records images on transparent plastic strips by means of photosensitive chemicals.

C. Editing the audio and video raw materials of a production to create a finished program.

D. A display of two different images at the same time.

E. To record film or video; also, an informal term for the production phase of a film or video project.

F. A transition in which the last image in the outgoing section gradually evolves into the first image of the incoming section.

G. The range of brightness values in an image, from black to white.

H. Separate sound recordings that are combined to create a program's sound track.

I. The ability to evaluate the content of visual media through an understanding of the way in which it is recorded and presented.

J. A copy of the original camera film that is edited to create a program.

K. A group of related scenes, like a chapter in a verbal composition.

L. The distribution of TV programs through electrical signals sent through the air.

M. The amount of fine detail carried by an image.

N. To record images and sounds as numerical data, either directly in a camcorder or in the process of importing them to a computer.

O. The process of actually videotaping the material for a program.

P. Video whose images show much finer detail than those of traditional video.

Q. Studio-based, multicamera video that is often produced and transmitted "live."

R. Preparing the program content and organizing the shoot, before production actually starts.

S. An imaginary world behind the video screen that looks like the real one but operates by quite different rules.

T. Recorded and (usually) transmitted for display continuously and in real time.

U. An audiovisual medium that records on magnetic tape or disk by electronic means; also, single-camera taped program creation in the manner of film production, rather than studio television.

V. A single video picture, like a single word in speech.

W. A group of closely related shots, like a verbal paragraph.

X. An appliance that both captures moving images (camera) and stores them on tape (recorder).

Y. A set of continuous images, comparable to a verbal sentence.

Chapter Quiz

Write the answer to each question below in the space provided.

1. Video combines elements of two older media: _____ and _____.

2. Name any two of the ways in which film is considered superior to video.

   _____

   _____

3. Name any two of the ways in which video is considered superior to film.

   _____

   _____

4. Why is it important to recognize the difference between "reality" in a video program and reality in the actual world?

   _____

   _____

5. Give one example of ways in which you can change time and space in the video world.

   _____

   _____

6. Name the two shortest of the four main units in the language of video expression.

   _____

   _____

Activity 1-1

The Literate Viewer: Keeping a Program Log

To determine what you enjoy watching on TV, try keeping a program log. To do this, simply fill in the information requested on the form below. Choose one night on which you usually watch two or more programs and keep the log for that night. (The sample line shows how to do this.) Do not worry if any of your program choices seems trivial or a waste of time—you will not put your name on this list. When your list has been completed, you and your classmates will analyze them as directed on the other side of this page.

Start Time	End Time	Length (Minutes)	Channel Name	Program Name
9pm	10pm	60	Nickelodeon	Daffy Duck (sample)

Total Minutes [   ]

Activity 1-1, continued

To analyze the TV watching habits of the class:

1. Gather up all the papers (for anonymity), then divide the class into small teams. Divide the stack of papers among the teams. Each team does the following:
2. Using one copy of this sheet, write down the information collected from the TV logs.
3. On a another copy of this sheet, compile the information from each of the team summary sheets.
4. Discuss the results. Which are the most popular programs? Why?
5. What is the average amount of TV time per person in the class? (Add up the number of hours watched, then divide by the number of people in the entire group.)

Program Name	Number Who Watched...	Length (minutes)	Total (minutes)
Daffy Duck (sample)	HHt ///	60	480

Total hours for all programs [ ] Divided by total number in a group [ ] = Average TV hours per person [ ]

Activity 1-2

Creating a Storyboard

Storyboards are so important in planning video programs that you need to be comfortable drawing them. Later activities in this workbook will show you how to do so—even if you think you have no graphic skills. The idea is to get you started by designing a storyboard for the very short sequence below.

As you draw your board, keep a few simple tips in mind:

- Draw one frame for each shot in your program. You should need all six frames.
- Use stick figures for people and circles for heads. Do not try to make finished drawings. The idea is to work quickly.
- Try to vary your shots, alternating wider views that include more of the scene with closer views that reveal more details.
- Below each frame, describe the action very briefly (for example, "runs down hall").

Here is the sequence. (Although pronouns are used for convenience, this sequence is not gender-specific. The character you draw can be female or male.)

Shortly after leaving a classroom, she realizes that she has left her backpack behind. Worried about losing it, she hurries back to the classroom. She looks all around the area where she was sitting, but the backpack is not there. Then she sees it sitting on the instructor's desk. Checking the backpack's contents, she is relieved to find everything there.

Getting Started

Reading Review

Write a brief answer for each question below. If you are not sure of an answer, review the appropriate section of the chapter to find it.

1. What is the difference between the power switch and the record switch?

   _____

   _____

2. What does the white balance control do?

   _____

3. What is the difference between *autofocus* and *autoexposure*?

   _____

   _____

4. How do you prepare a new videotape for recording?

   _____

   _____

5. What is the difference between *panning* and *tilting* the camera?

   _____

6. What does *default* mean?

   _____

7. Which controls are set in automatic mode when you turn the camcorder on?

   _____

8. How do you set up a tripod for use?

   _____

   _____

9. Name three common problems to avoid in recording each shot.

   _____

   _____

   _____

10. What is *head room*?

    _____

11. What is the difference between *look room* and *lead room*?

    _____

    _____

12. What is the *rule of thirds* and how can you use it?

_____

_____

_____

13. How should a new camera angle be different from the previous one?

_____

_____

14. What are jump cuts and why should you avoid them?

_____

_____

15. Why should you place the camcorder so that the microphone is close to the subject?

_____

_____

16. Why should you move subjects away from background noise like traffic?

_____

Vocabulary Review

Match each term to its definition by writing the letter of the correct definition in the space provided. (Not every one of these terms is defined in the *Technical Terms* section in the textbook, and some of the definitions do not apply to any term listed here.)

Terms

_____ 1. Default

_____ 2. Jump cut

_____ 3. Autofocus

_____ 4. Background noise

_____ 5. Pan

_____ 6. Camera angles

_____ 7. Zoom

_____ 8. Record switch

_____ 9. Head room

_____ 10. Lead room

_____ 11. Tripod

_____ 12. Look room

_____ 13. Power switch

_____ 14. Camcorder

_____ 15. Roll to raw stock

_____ 16. Lens cap

_____ 17. Rule of thirds

_____ 18. Standby

_____ 19. White balance

_____ 20. Autoexposure

Definitions

A. The distance between the subject and the edge of the frame toward which the subject is moving.

B. To pivot the camcorder horizontally on its support.

C. An appliance that both captures moving images and stores them on tape.

D. The most common digital recording format.

E. The control that turns the camcorder on and off.

F. A condition in which the camera is on and operating but not recording.

G. The camera system that ensures that the subject of the image appears clear and sharp.

H. The camera system that delivers the correct amount of illumination to the recording mechanism, regardless of the light level of the shooting environment.

I. The amount of power remaining in a camcorder battery.

J. An aid to composition in the form of an imaginary tic-tac-toe grid superimposed on the image. Important picture components are aligned with the lines and intersections of the grid.

K. To magnify or reduce the size of a subject by changing the focal length of the lens while recording.

L. The camera system that neutralizes the color tints of different light sources, such as sunshine and halogen lamps.

M. A setting that is selected automatically unless the user changes it manually.

N. The part of the lens that changes the exposure.

O. The distance between the subject and the edge of the frame toward which the subject is looking.

P. The positions from which the camcorder records shots.

Q. Two shots edited together that are too similar.

R. A protective cover for the camcorder lens.

S. A lens setting that brings distant subjects close.

T. To advance a videotape through previously recorded sections to blank tape, in preparation for additional recording.

U. A method of composition that divides the screen in half both horizontally and vertically.

V. The control that changes the camcorder from standby to record mode, and back again.

W. A three-legged camera support.

X. The distance between the top of a subject's head and the upper edge of the frame.

Y. General sounds in the recording environment.

Chapter Quiz

Answer each question below. For True/False or Multiple Choice questions, circle the correct answer. When more than one answer seems reasonable, choose the best one. For other questions, write the answer in the space provided.

1. The white balance switch should be left in which position?
    A. outdoor
    B. automatic
    C. indoor
    D. fluorescent

T F 2. A default setting is a camcorder function control set to automatic.

3. How do you prepare a new tape for use?

_____

_____

4. Panning means moving the camcorder _____.
    A. up and down
    B. back and forth
    C. both

5. Name three things to check in the viewfinder before shooting.

_____

_____

_____

6. Which three controls should be set to automatic?

_____

_____

7. Name two ways to avoid shaky handheld shots.

_____

_____

T F 8. You should set the tripod up with one leg facing you, and straddle it to shoot.

9. Name two common problems with shots.

_____

_____

10. Keeping the subject's eyes in the top third of the frame helps maintain enough _____.
    A. head room
    B. look room
    C. lead room

T F 11. Look room applies only to human subjects.

T F 12. The rule of thirds helps avoid jump cuts.

13. To use the rule of thirds, align important composition elements with

_____

14. On-camera microphones record best when they are _____.
    A. turned up in volume
    B. close to the subject
    C. placed level with the subject's mouth

Activity 2-1

Storyboard Practice

To draw storyboard heads for closeups, you need to *aim* them and provide them with *expressions*. You also need to indicate *gender,* where appropriate. To practice these skills, try copying each of the heads on this page and the next one in the empty frame directly under it. As you work on your sketches, notice the various suggestions printed on the worksheet. Remember: try for clean outlines rather than finished drawings. The simpler the sketch, the better. (For a sheet of blank frames to practice with, turn to the last page in this chapter).

Front/smiling. Eyes near middle of head, straight line nose.

3/4 front/worried. Nose, ear, and shoulder show face turned somewhat.

Profile/angry. Mouth and eyebrow(s) show anger.

3/4 rear. For profile and rear views of heads, use hair and ear(s) to define.

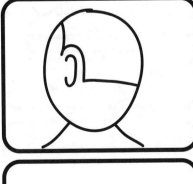

Looking up/neutral emotion. Features high on face, chin shows.

Looking down/unhappy. Features low on face, hair prominent.

Activity 2-1, continued

Use stick figures for full-length characters. Express postures with the angles of the shoulders and hips, plus the curve of the spine. For female figures, add long hair, and, if appropriate, a skirt from the hip line down. Note the arrows used to indicate the direction of movement. (You can omit them, if you prefer, as you copy the models.)

Front/female/scheming or suspicious. Hair defines gender.

Profile/female/annoyed. Eyebrows and mouth show feelings.

Standing figure. Shoulders, torso, and hips are straight lines.

Sitting figure. Shoulders and hips match diagonal seat.

3/4 view walking figure. Diagonal shoulders/hips show body turn.

Profile, running figure.

Activity 2-2

The Literate Viewer: Counting Commercials

How many commercials are there, anyway?

To find out, videotape one hour of evening programming and analyze the results. Here is how to do it.

1. Choose any broadcast or cable channel that you enjoy watching. Prepare to tape it with your VCR as you would any other program.
2. Exactly on the hour (for example, 8:00 p.m.), begin recording. Continue recording for slightly more than one hour, to be sure that you include the starting point of the next program.

At a convenient later time, you can analyze the commercials:

1. Cue the tape in the VCR to the exact start of the first actual program material (not commercials or "promos") after the beginning of the tape. Set the tape time counter to zero.
2. Using the worksheet below, go through the tape, logging only the materials that are *not* part of a program (commercials or "promos" for news or other shows.)

The first line is a sample. When you reach the start of the program beginning exactly one hour later, stop logging. To find the total minutes in an hour that are not actual program, add up the seconds listed on your log, then divide by 60 (to find the number of minutes). Subtract your total from 60 minutes to identify the actual program time.

Product or Promo	Seconds	Product or Promo	Seconds
10 o'clock news (sample)	20		
		Total nonprogram minutes:	
		Total program:	

Practice Frames

Video Space

Reading Review

Write a brief answer for each question below. If you are not sure of an answer, review the appropriate section of the chapter to find it.

1. What are the four laws of video space?

_____

_____

_____

_____

2. What is the difference between "framing" something and "framing off" something?

_____

3. Give two examples of using the frame to control content.

_____

_____

4. Give two other examples of the power of the frame.

_____

_____

5. What two clues do viewers use to judge apparent movement on the screen?

_____

_____

6. Explain one way to make a model spaceship move across the screen.

_____

7. In the world on the screen, what determines "vertical" (up and down)?

_____

8. Name the three dimensions in the video world.

_____

9. Explain how to make a subject appear to float horizontally above the sidewalk.

_____

_____

10. Describe how to visually shorten a long walk.

_____

_____

11. Explain how to create the shot of a tiny person standing on another person's hand, as shown at the beginning of Chapter 3 in the textbook.

_____

_____

12. In the "wall-climbing" setup in Chapter 3, why does one crew member stand on the front of the line and a second crew member hold up the back of the line?

_____

_____

Vocabulary Review

Match each term to its definition by writing the letter of the correct definition in the space provided. (Not every one of these terms is defined in the *Technical Terms* section in the textbook, and some of the definitions do not apply to any term listed here.)

Terms

_____ 1. Sell

_____ 2. Frame off

_____ 3. Gag

_____ 4. Frame

_____ 5. Screen direction

_____ 6. Cheat

_____ 7. Setup

_____ 8. Height

_____ 9. To frame

_____ 10. Breadth

Definitions

A. The video dimension parallel to the top and bottom of the screen.

B. To exclude an object completely from the screen.

C. To move a subject from its original place (to facilitate another shot) in a way that is undetectable to the viewer.

D. Something that appears real on the screen, but is actually a trick.

E. The apparent positions of objects on the screen, with respect to one another.

F. The apparent movement of subjects in the frame.

G. The video dimension that appears to extend from front to back on the screen.

H. The four edges that make up the border of the screen. Also, a single video image.

I. To add details that increase the believability of a screen illusion.

J. The purposeful arrangement of the parts of an image.

K. The video dimension parallel to the sides of the frame.

L. To feature a subject in a composition.

M. Any effect, trick, or stunt in a movie.

N. The orientation of on-screen movement with respect to the left and right edges of the frame.

O. A single camera position, usually including lights and microphone placements as well.

Chapter Quiz

Answer each question below. For True/False or Multiple Choice questions, circle the correct answer. When more than one answer seems reasonable, choose the best one. For other questions, write the answer in the space provided.

1. What is outside the frame does not exist, unless

   _____.

2. Height and breadth are determined solely by

   _____.

3. Four things in video space are not *fixed* (always the same). Name two of them.

   _____

   _____

4. Direction is determined solely by

   _____.

5. Which of the following is *not* a meaning of the word "frame?"
   A. The border around the video image.
   B. An element of visual composition.
   C. A single video image, one of 30 per second.

T F 6. The frame of traditional video is proportioned 3:2.

7. "Framing off" something means

   _____.

8. Composition is _____.
   A. the organization of the elements (parts) of a picture.
   B. the creation of the illusion of depth on a flat screen.

9. The three dimensions in the video world are

   _____.

T F 10. In the real world, depth is only an illusion.

11. An on-screen effect created by a trick is called a

   _____.

T F 12. "Cheating" means placing the camera far from the subject and compensating with a telephoto lens.

13. "Controlling spatial relations" means _____.
   A. selecting wide angle or telephoto lenses.
   B. dealing with extraterrestrial aunts and uncles.
   C. putting things together on-screen that are not together in reality.

14. A single camera position is called

   _____.

T F 15. Real-world movements in different directions can seem like the same direction on the screen.

16. In conventional screen direction, on-screen "left" and "right" are based on

_____

_____ .

17. In conventional shots, planes flying from New York to London are shown traveling from

_____ to _____ .

Activity 3-1

Storyboard Practice

On this page and the next are some drawings that are frequently used in making storyboards. To practice them, try copying each sketch in the empty frame directly under it. As you work on your sketches, notice the various suggestions printed on the worksheet. Try for clean outlines, rather than finished drawings. The simpler the sketch, the better.

Often, you can indicate camera height by drawing a horizon line. Notice how simple shapes suggest subjects and arrows indicate movement.

High angle: desert.   Neutral angle: football field.   Low angle: street.

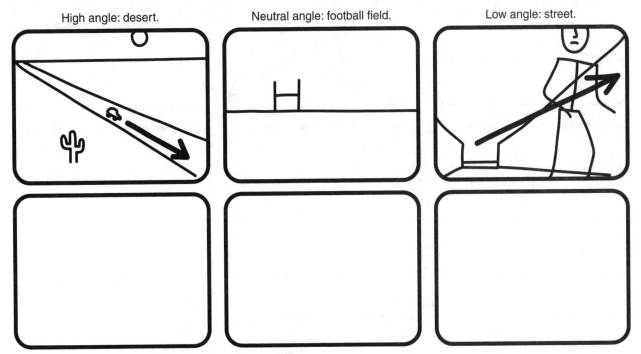

Camera Pan
The two frames show the start and end of the move. Sometimes, the drawing continues between frames. Arrow indicates movement.

Zoom in. Small frame equals end composition.

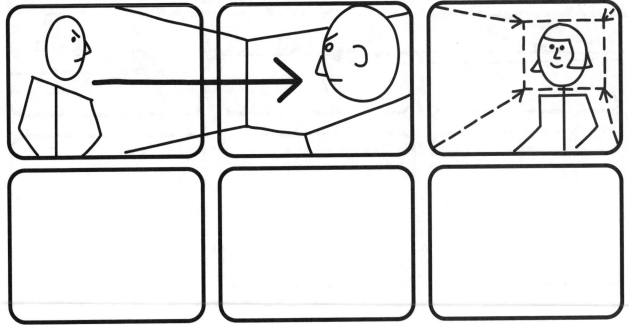

Here are some easy ways to show spatial relationships in storyboard sketches. Notice the free use of arrows and the punctuation marks employed to show emotions.

Over-the-shoulder two-shot. The nearer person overlaps the more distant one. Nearer person's mouth and eyes are not visible.

A "glance-object pair." The arrows reinforce the idea that the subject is looking at…

a bomb.

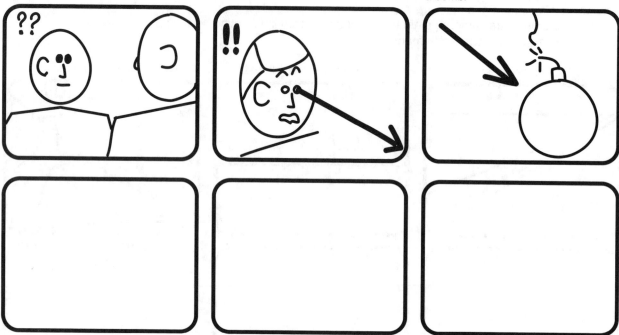

Indicating Directions

The same subject is shown twice, with an arrow indicating that he reverses direction.

Two subjects come from different directions and walk together.

Depth

In this chase scene, the closer subject is larger. The road and tree add extra depth.

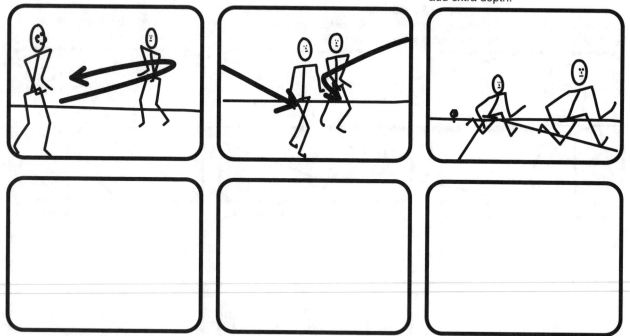

Activity 3-2

A Story to Board

Turn the story below into a storyboard. (As always, the pronouns are only for convenience; the story is not gender-specific.)

1. Number your frames. If you cross out an unneeded frame, do not renumber the remaining frames. If you wish to replace a frame, cross it out, draw the replacement on the next page, and number it to match the original.

2. If you need more frames, use the blanks on the next page.

Waking up, she looks at her alarm clock. Forty-five minutes late! She goes through drawers and closet like a tornado, looking for clothes. She gulps a cup of cold coffee from yesterday's brew. She races to her car and tries to start it: The battery is dead. Desperately, she takes a bus, which is crowded and lets her off a long way from campus. She runs the last several blocks, arriving disheveled and winded. With her last energy, she runs up the stairs and down the hall to the classroom door, where she finds several other students. Asking why they're still outside, she learns that the change from daylight saving time to standard time was the day before, and she's actually early.

Practice Frames

 **Video Time**

Reading Review

Write a brief answer for each question below. If you are not sure of an answer, review the appropriate section of the chapter to find it.

1. What are the four aspects (characteristics) of time that you can control in the video world?

   _____

2. In which two areas can you alter the *speed* of video time?

   _____

3. What are the effects of slight fast motion, moderate fast motion, and extreme fast motion?

   _____

   _____

4. How can you shorten or lengthen video time in editing?

   _____

   _____

5. What is serial time flow?

   _____

6. What is parallel time flow?

   _____

7. What are two ways to make video time run backward?

   _____

   _____

8. What does "time coherence" mean with regard to editing video and audio?

   _____

9. What is a "split edit?"

   _____

10. What happens in a split edit when audio leads?

   _____

11. What happens in a split edit when video leads?

   _____

12. How can voice-over narration separate the video and audio time flows?

   _____

Vocabulary Review

Match each term to its definition by writing the letter of the correct definition in the space provided. (Not every one of these terms is defined in the *Technical Terms* section in the textbook, and some of the definitions do not apply to any term listed here.)

Terms

_____ 1. Audio leads

_____ 2. Screen time

_____ 3. Fast motion

_____ 4. Split edit

_____ 5. Cross cutting

_____ 6. Flashback

_____ 7. Overlapping action

_____ 8. Serial time flow

_____ 9. Parallel time flow

_____ 10. Omitting action

Definitions

A. A video time speed, within a shot, that is slower than that of the real world.

B. The length of real-world time required to display a piece of video.

C. A sequence that takes place earlier in the story than the sequence that precedes it.

D. A series of events moving forward in a single sequence.

E. The rate at which time passes within a video shot.

F. Repeating the action from the end of the outgoing shot at the start of the incoming shot, to lengthen the screen time of the scene.

G. An edit in which the audio and video of the new shot do not begin simultaneously.

H. Speech on the sound track, spoken by someone who does not appear on screen.

I. A video time speed, within a shot, that is faster than that of the real world.

J. The speed at which time seems to be passing, as displayed in a succession of edited shots.

K. Two of series of events moving forward in separate sequences, presented alternately in parts by cross cutting.

L. Showing two actions at once by alternating back and forth between them, presenting part of one, then part of the other, then back to the first, and so on.

M. Cutting the action both at the end of the outgoing shot and at the start of the incoming shot, to shorten the screen time it takes.

N. A split edit in which the sound from the preceding shot continues over the visual of new shot.

O. A split edit in which the sound from the new shot begins over the end of the preceding one.

Chapter Quiz

Answer each question below. For True/False or Multiple Choice questions, circle the correct answer. When more than one answer seems reasonable, choose the best one. For other questions, write the answer in the space provided.

1. Name two of the four characteristics of video time that you can control.

   _____

   _____

2. Circle the two areas in which you can change the speed of video time.
   A. Within shots.
   B. Between sequences.
   C. Between shots.

T   F   3. Moderate slow motion makes action look comical.

4. What does "overlapping action" mean?

   _____

   _____

5. An editor can shorten time between shots by

   _____

   _____ .

6. Cutting back and forth among two or more actions creates _____ time flow.

7. List two ways to make video time run backward.

   _____

   _____

T   F   8. When video leads in a split edit, the picture begins a new shot before the sound does.

Activity 4-1

Mini-project: Magic Pushups

Description

By turning the camcorder sideways, you can make a subject appear to float above the ground.

Objective

To show that "up" and "down" are determined by the frame.

Materials

- A foam drinking cup.
- Some chewing gum.
- A subject to do pushups.
- A camcorder kit.

Preproduction

Plan the sequence by storyboarding it, using the blank storyboard form on the next page. Locate a concrete floor and a concrete wall that appear similar (they need not be in the same location). Shoot in overcast weather or select a time of day to avoid throwing obvious shadows. Follow these production guidelines:

Production

1. At the floor location, tape:
   - An establishing shot of the subject carrying the cup to the pushup location, drinking from the cup, and setting it down.
   - The subject getting into position and preparing to do pushups.
   - The subject at the end of a pushup, sitting up, drinking more water, then rising and leaving.

2. At the wall location:
   - Turn the camera 90 degrees so that the frame top and bottom are vertical.
   - Position the subject to match the screen direction of the floor location shots. Frame the top half of the subject's body.
   - Use chewing gum to fix the cup bottom to the wall. Make sure that *in the frame*, the cup appears in the same relationship to the subject as in the floor location shots.

3. Tape the following action:
   - The subject does two or three "difficult" pushups.
   - The subject does a few one-handed pushups, then one-finger pushups.
   - The subject lifts the last finger and floats in the air.
   - The subject slowly returns to the "floor."

Postproduction

Edit the shots together, beginning with the opening "floor" shots, then the "wall" shots, and finally the closing "floor" shots.

Discussion

1. Why does the pushup gag work?

2. What is the purpose of the water cup and drinking the water?

3. How does matching action from shot to shot help sell the gag?

4. What are some problems (hint: such as light and clothing) that could (or did) reveal the trick?

Activity 4-2

Mini-project: It Came from Outer Space!

Description

By manipulating scale and distance, you can turn a pair of pie plates into a giant flying saucer.

Objective

To show that in the video world, size and distance are controlled by you.

Materials

- Two disposable aluminum pie plates.
- Stapler.
- Thin white thread or clear nylon fishing leader.
- Stick or pole at least 4 feet long.
- A camcorder kit.

Preproduction

1. To construct your flying saucer, first find the exact center of one pie plate. Punch a small hole in it, and attach several feet of thread through the hole by securing it on the inside with tape. Set the first pan, bottom-up, on the second pan and staple their edges together in several places.
2. Tie the other end of the thread to one end of the pole. *Hint:* Make a temporary knot so that the thread length can be adjusted later, as needed.
3. Create a storyboard sequence in which a giant alien ship flies in and menaces the people on the ground. Shots might include:
   - The saucer appearing in the sky over local buildings.
   - People reacting fearfully to the invasion.
   - The saucer hovering over a victim and making him disappear.
   - The saucer chasing terrified people.
   - The saucer landing and aliens (you think them up) emerging from its far side.

Plan the sequence by storyboarding it on the next page.

Production

The trick is to swing the pie plates through the frame near enough to the camera to make them seem huge in comparison with people and buildings much farther away. To help keep the pie plates in focus, shoot outdoors in bright light, using a wide angle lens setting. If possible, the saucer puppeteer should be able to watch the camcorder's external view screen while manipulating the pie plates. To make the victim disappear, align him under the nonmoving saucer, with the camera locked down. After a few seconds of shooting, have the victim move out of frame, leaving only the saucer. Continue shooting for a few more seconds. For the emerging "aliens," do the trick in reverse: "Land" the saucer in front of an empty space, then have the "aliens" move behind it and, after a few seconds, walk around its front while the shot continues.

Postproduction

Edit the shots, alternating the saucer with human reactions. To "vanish" the victim, cut out the footage in which he walks out of frame, then dissolve the beginning of the shot into the end of it. Treat the emerging aliens the same way.

Discussion

1. What does this project demonstrate about size in the video world? What about depth (distance from the camera)?
2. Why is focus a critical problem in getting successful footage?

Extra

If a blue or green compositing screen is available, shoot moving shots of the saucer to matte into live action exteriors. To make the saucer "move" across the screen, begin with it hanging in the center of the blue screen, out of frame on the "start" side, then pan the camcorder across it until it disappears off the "end" side. To make the ship recede or advance, hold it still while the camera zooms out or in. To make it "land" or "take off," suspend it by one edge, bottom toward the camcorder and zoom in or out. *Hint*: At the zoom-in position, the saucer should fill the whole frame. (Terrible crunching sounds on the audio track will help sell the gag.)

Video Composition

Reading Review

Write a brief answer for each question below. If you are not sure of an answer, review the appropriate section of the chapter to find it.

1. What is "video composition?"

_____

2. How does composition help viewers "decode" an image?

_____

3. What is "simplicity" in composition?

_____

4. What is "order" in composition?

_____

5. What is "balance" in composition?

_____

6. Why is the frame around the image so important to composition?

_____

_____

7. In addition to defining area, the frame establishes _____ dimensions.

_____

8. What is "emphasis" in composition?

_____

9. How does "significance" create emphasis in a composition?

_____

10. How does selective focus create emphasis in a composition?

_____

_____

11. What is the meaning of "perspective" as used in video?

_____

12. How does "apparent size" suggest distance?

_____

13. How does "overlap" suggest distance?

_____

14. How does "convergence" suggest distance?

_____

15. How does "vertical position" suggest distance?

_____

16. How does "sharpness" suggest distance?

_____

17. How does "color intensity" suggest distance?

_____

18. Name two compositional techniques you can use to direct the viewer's attention.

_____

19. Explain the "rule of thirds."

_____

_____

Vocabulary Review

Match each term to its definition by writing the letter of the correct definition in the space provided. (Not every one of these terms is defined in the *Technical Terms* section in the textbook, and some of the definitions do not apply to any term listed here.)

Terms

_____ 1. Tilting

_____ 2. Overlap

_____ 3. Contrast

_____ 4. Asymmetrical balance

_____ 5. Focus

_____ 6. Letterboxed image

_____ 7. Picture plane

_____ 8. Staging in depth

_____ 9. Widescreen video

_____ 10. Rule of thirds

_____ 11. Panning

_____ 12. Leading lines

_____ 13. Brightness

_____ 14. Leading the eye

_____ 15. Perspective

_____ 16. Vertical position

_____ 17. Balance

_____ 18. Emphasis

_____ 19. Receding lines

_____ 20. Composition

Definitions

A. The purposeful arrangement of elements in an image.

B. The simulation of depth in a two-dimensional image.

C. Lines on the picture plane that emphasize an element by pointing to it.

D. A method of composition that aligns important visual elements with the lines and intersections of a tic-tac-toe grid.

E. The actual two-dimensional image.

F. A composition in which dissimilar elements have equal "visual weight."

G. Rotating the camcorder vertically.

H. A wide screen image displayed in the center of a regular TV screen.

I. Video using a screen proportioned 16 to 9.

J. The position of a pictorial element on a scale from black to white.

K. A form of perspective based on the height of an element on the picture plane.

L. Using a contrasting color to attract the eye.

M. Moving the camera horizontally.

N. Diagonal lines in a composition that enhance its apparent depth.

O. That part of the image which appears sharp and clear.

P. Enhancing apparent depth by placing some pictorial elements in front of others.

Q. Using compositional techniques to direct the viewer's attention.

R. Positioning subjects and camcorder to exploit perspective in the image.

S. Distributing objects to create equal "weight" in different parts of the image.

T. The difference between one pictorial element and others; also, the ratio of the brightest part of an image to the darkest.

U. The process of calling attention to a pictorial element.

V. Composition in which visual elements are evenly placed and opposed.

W. Method of composition that divides the image into halves.

X. Pictorial element that attracts the eye because it means something special to the viewer.

Y. Identifying and understanding the elements in a composition.

Chapter Quiz

Answer each question below. For True/False or Multiple Choice questions, circle the correct answer. When more than one answer seems reasonable, choose the best one. For other questions, write the answer in the space provided.

1. Visual composition is the _____ arrangement of the components of a visual image.
   A. pleasing
   B. purposeful
   C. artistic
   D. logical

2. To help viewers decode (understand) an image it should be organized according to three principles:

   _____

   _____.

3. Why is the frame so important in composition?

   _____

   _____

T F 4. Emphasis is a method of simulating depth in a composition.

T F 5. Contrast is a method of emphasizing one element in a picture by making it look different from the other elements.

6. Perspective may be defined as

   _____

   _____.

7. The six techniques used to create perspective are (in any order):

   _____

   _____.

8. To use perspective in composing images, you can _____.
   A. Incorporate perspective elements that naturally occur in the shot.
   B. Arrange pictorial elements to emphasize apparent depth.
   C. Both.

T F 9. In any composition, the biggest pictorial element is always the most important.

10. Organizing pictorial elements so that their characteristics are different and their positions are not evenly opposed is called _____.

11. A "rule of thirds" composition grid looks like _____.
    A. a checkerboard
    B. a tic-tac-toe game
    C. cross-hairs in a gun sight

12. If you use one vertical and one horizontal line to divide an image into four equal parts, the most visually powerful part is _____.

13. What do leading lines do in a composition?

   _____

   _____

Name _____

Date _____

Activity 5-1

Understanding Composition

Here are nine compositions. Most, *but not all,* are based on the "rule of thirds." On each, draw the lines from the tic-tac-toe grid that apply to this picture. *Hint:* these pictures do not necessarily need all four grid lines, as the example demonstrates. On the pictures that do not use the rule of thirds guide, try drawing lines indicating the main components of the composition. See the examples in Chapter 5 of the textbook.

Activity 5-2

Types of Perspective

Here are six compositions that use one or more types of perspective to achieve the appearance of depth. Under each picture, write the type(s) of perspective used. Perspective types include *size, overlap, convergence, vertical position, sharpness,* and *color intensity*.

In this example, perspective types used include *size, convergence,* and *vertical position*.

_____

_____

_____

_____

_____

_____

Activity 5-3

Viewfinder

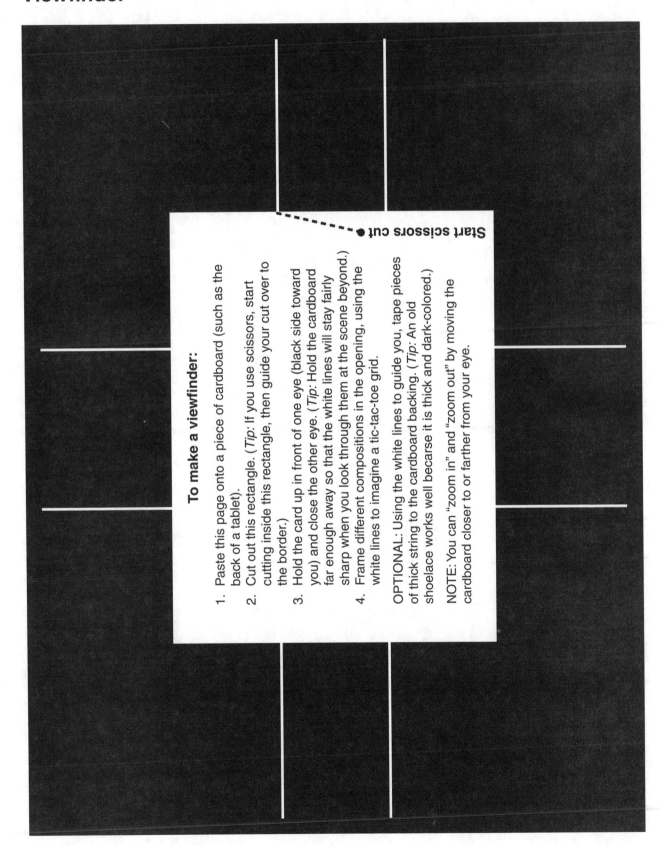

Start scissors cut

To make a viewfinder:

1. Paste this page onto a piece of cardboard (such as the back of a tablet).

2. Cut out this rectangle. (*Tip:* If you use scissors, start cutting inside this rectangle, then guide your cut over to the border.)

3. Hold the card up in front of one eye (black side toward you) and close the other eye. (*Tip:* Hold the cardboard far enough away so that the white lines will stay fairly sharp when you look through them at the scene beyond.)

4. Frame different compositions in the opening, using the white lines to imagine a tic-tac-toe grid.

OPTIONAL: Using the white lines to guide you, tape pieces of thick string to the cardboard backing. (*Tip:* An old shoelace works well becarse it is thick and dark-colored.)

NOTE: You can "zoom in" and "zoom out" by moving the cardboard closer to or farther from your eye.

Back of viewfinder sheet. Paste this side to cardboard backing.

Name _____

Date _____

Video Language

Reading Review

Write a brief answer for each question below. If you are not sure of an answer, review the appropriate section of the chapter to find it.

1. Define the video term "shot."

   _____

2. If an image is like a word in verbal language, what units in video language correspond to verbal sentences, paragraphs, and chapters?

   _____

3. What is the definition of a camera angle?

   _____

4. Give two examples of camera angles named by subject distance.

   _____

5. Give two examples of camera angles named by horizontal angle.

   _____

6. Give two examples of camera angles named by height.

   _____

7. Give two examples of camera angles named by lens perspective.

   _____

8. Give two examples of camera angles named by shot purpose.

   _____

9. Give two examples of camera angles named by shot population.

   _____

10. Explain the term "matching action."

    _____

11. Describe the difference in appearance between a fade and a dissolve.

    _____
    _____

12. Explain why digital video effects (DVEs) are so widely used in commercials and music videos.

    _____

Vocabulary Review

Match each term to its definition by writing the letter of the correct definition in the space provided. (Not every one of these terms is defined in the *Technical Terms* section in the textbook, and some of the definitions do not apply to any term listed here.)

Terms

_____ 1. Program

_____ 2. Cut together

_____ 3. Wide angle lens

_____ 4. Dissolve

_____ 5. Fade-out

_____ 6. Wipe

_____ 7. Jump cut

_____ 8. Image

_____ 9. Frame

_____ 10. Point of view

_____ 11. Act

_____ 12. Setup

_____ 13. Shot

_____ 14. Scene

_____ 15. Camera angle

_____ 16. Match points

_____ 17. Continuity

_____ 18. Digital video effect (DVE)

_____ 19. Sequence

_____ 20. Telephoto lens

Definitions

A. An edit in which the incoming shot is too similar to the outgoing shot.

B. To follow one shot with another.

C. A longer segment of program content, usually consisting of several related scenes.

D. A lens or a setting on a zoom lens that minimizes subjects and magnifies apparent depth by filling the frame with a wide angle of view.

E. A DVE in which the screen seems to revolve to show its other side.

F. A fade-in that coincides with a fade-out, so that the incoming shot gradually replaces the outgoing shot.

G. A major section (usually between 10 and 45 minutes) of a longer program.

H. A transition in which the image begins at full brightness and gradually darkens to pure black.

I. A camera angle that frames the subject from head to waist.

J. The organization of video material into a coherent presentation.

K. A single set of visual information.

L. The places, in two shots, where they can be cut together to make the action appear continuous.

M. A vantage point from which the camera records a shot.

N. A transition in which the image begins as pure black and gradually lightens to full brightness.

O. An arrangement of production equipment placed to record shots from a certain point of view.

P. Any complete video presentation.

Q. The position from which a shot is taken.

R. A short segment of program content, usually made up of several related shots.

S. A very close shot designed to reveal small details.

T. A lens or a setting on a zoom lens setting that magnifies subjects and minimizes apparent depth by filling the frame with a narrow angle of view.

U. Any digitally created transitional device other than a fade or dissolve.

V. A single still picture, 30 of which make a second of NTSC video. Also, the border around the image.

W. An effect that signals the change from one sequence to the next.

X. A single continuous recording.

Y. A transition in which a line moves across the screen, covering the outgoing shot with the incoming shot.

Chapter Quiz

Answer each question below. For True/False or Multiple Choice questions, circle the correct answer. When more than one answer seems reasonable, choose the best one. For other questions, write the answer in the space provided.

1. The individual pieces of video are called _____.

2. Collectively, the rules governing the organization of shots are called _____.

T   F   3. Compared to verbal language, a *shot* is like a *word*.

4. A camera angle frames a subject from a particular _____.
   A. position
   B. image size
   C. Both.

5. In a full shot, a standing subject occupies the frame from _____ to

   _____.

6. A medium shot cuts the subject off at _____.

T   F   7. In a closeup, the subject fills the frame from below the chin to the forehead.

8. Match the terms to their descriptions:

   _____ Horizontal angle   A. The location of the camera on a vertical arc.

   _____ Camera height   B. The way a particular lens renders an image.

   _____ Lens perspective   C. The number of subjects in the shot.

   _____ Shot purpose   D. The camera location on a circle around the subject.

   _____ Shot population   E. The intended use of the shot in the program.

9. A telephoto lens makes subject seem _____.
   A. closer
   B. farther away

T   F   10. Technically, the only difference between various lens focal lengths is the angle of view.

11. Explain either one of the two ways to match action from shot to shot.

   _____

   _____

12. Almost all programs start with a _____ and end with a

   _____.

13. The letters DVE stand for.

   _____

14. A moving line that changes one image into another is called a _____.

15. Which component of a program is longer, a scene or a sequence?

   _____

16. A single, uninterrupted recording is called _____.

Name_____

Date_____

Activity 6-1

Mini-project: Talking to Yourself

Description

This activity allows you to manipulate video time, so that a subject can have a conversation with him/herself.

Objective

To show how completely video time is controlled by the program maker.

Materials

- A camcorder kit.
- Two subjects.
- A few lines of dialogue. (See the back of this page for suggestions.)
- A quiet location (for quality audio) and a fairly simple background.

Preproduction

Cast two people as Subjects A and B. Have them rehearse the selected sequence (or write one of your own). Then, have them switch roles and rehearse it again.

Production

Place the subjects standing facing each other and frame them with a medium (waist-level) two-shot. Make sure that the camcorder and tripod are firmly locked down. Tape the scene in one shot. Then have Subjects A and B change places and switch roles. Tape the scene a second time.

Postproduction

Using a 50/50 split screen (a soft-edge wipe works well), synchronize the first and second recordings so that the same subject is in the left half in one and in the right half in the other. Render the shot so that either A or B is playing both parts.

To achieve a variety of different effects:

1. Edit the two shots again, this time with subject B playing both parts.
2. Synchronize the two different versions as A and B rolls.
3. Dissolve back and forth during the shot so that A morphs into B and back to A again.

Discussion

1. Analyzing the finished shots, how could the illusion have been made more convincing?
2. How could you achieve the same effect by editing two different shots together?
3. What are some other effects you could achieve using this split screen technique?

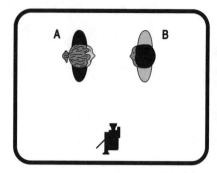 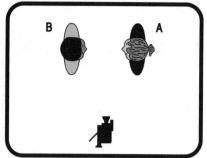

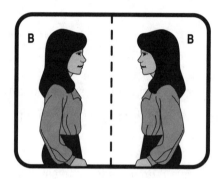

Suggested Scripts

Here are some simple texts to use in creating the effect. (It makes editing easier if A and B maintain the same pace and rhythm.)

The first is a short scene.

A: Hey! How's it going?

B: Not so bad. You?

A: Tell you the truth, I'm kind of spooked.

B: What's wrong?

A: I keep seeing people who look just like me.

B: No kidding! Do I look like you?

A: Exactly like me! Like my twin!

B: Hey! No need to get insulting! *(Walks out of frame.)*

The second sample is a nursery rhyme.

A: One, two—Buckle my shoe.

B: Three, four—Shut the door.

A: Five, six—Pick up sticks.

B: Seven, eight—Lay them straight.

A: Nine, ten—A great fat hen.

Activity 6-2

Mini-project: The Endless Staircase

Description

Through camera positions and editing, you can stretch a staircase to almost three times its real-world length.

Objective

To show how you control video time through editing.

Materials

- A camcorder kit.
- One subject.
- Staircase. (Review diagrams of the ground plan and camera angles for guidance in selecting a location.)

Preproduction

Find a staircase with at least eight or ten steps that allows the camera to shoot across it as well as from the top or bottom.

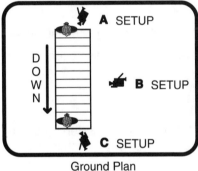

Ground Plan

Production

- Tape the subject ascending or descending the stairs completely from each of three angles: setups A, B, and C.
- Make sure that the camcorder is always at one side of the stairs and the subject is at the other.
- Compose setups A and C as wide shots; and setup B as a medium shot, framing off the steps.
- Record setup B twice: first, following the subject down the steps; then with the camera locked, framing the center part of the action, so that the subjects enters and exits the frame.

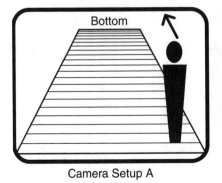

Camera Setup A

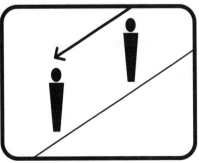

Camera Setup B

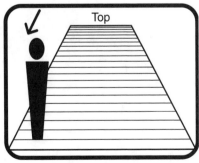

Camera Setup C

The number of steps will vary according to the stairs you choose. Also, the staircase you use may require shooting from the left side instead of the right.

Postproduction

Edit the three shots so that:

1. Setup A covers the action from steps one through seven.

2. Setup B covers the action from steps three through eight, with the camera following the subject.

3. Setup C covers the action from steps four through ten.

Shot 1: Steps 1 thru 7

Shot 2: Steps 3 thru 8

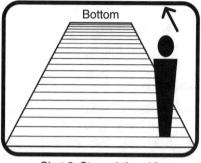

Shot 3: Steps 4 thru 10

4. Make a second assembly identical to the first; but this time, in Setup B, the subject enters and exits the shot. (Experiment by leaving different amounts of empty frame before and/or after the subject appears.)

5. Make a third assembly, using only setups A and C, omitting setup B completely.

6. Make a fourth assembly edited to make the staircase seem *shorter* rather than longer. (Try it both with and without setup B.)

Discussion

1. How does the "endless staircase" effect work with setup B as a moving shot? With setup B as a fixed-camera shot? In each case, why? What are some potential hazards of following the subject down the stairs in setup B?

2. How well does the sequence work without setup B? Why?

3. How could you reframe the setup A and C wide shots so that they would work better without the buffer of setup B?

4. Typically, an editor would shorten the staircase descent in order to condense unimportant action. When this is the case, is setup B necessary, or will viewers accept the jump between setups A and C? Why?

Video Sound

Reading Review

Write a brief answer for each question below. If you are not sure of an answer, review the appropriate section of the chapter to find it.

1. Name three ways through which program audio delivers information.

   _____

2. Explain how dialogue can easily deliver information about character and emotions.

   _____

3. Give two reasons why narration can deliver information efficiently.

   _____

4. List two roles of sound effects in conveying information.

   _____

5. How does sound communicate a sense of locale (environment)?

   _____

6. Why must the editor regulate sound volume and emphasize important sounds on the audio track?

   _____

7. How can sound help blend together shots made at separate times and in different places?

   _____

8. Name at least three types of audio elements used to make a sound track.

   _____

   _____

9. Aside from music, which type of sound is typically used to enhance mood?

   _____

10. Name two sound effects mentioned in the text and the feelings they evoke.

    _____

    _____

11. Give three functions of music in documentary and training programs.

    _____

    _____

    _____

Name_____

Date_____

Chapter Quiz

Answer each question below. For True/False or Multiple Choice questions, circle the correct answer. When more than one answer seems reasonable, choose the best one. For other questions, write the answer in the space provided.

1. The process of defining a project, organizing its content, and preparing a detailed production blueprint is called

_____.

2. Defining a program involves specifying these five criteria:

_____

_____.

3. Which of the following objectives is phrased in active terms?
   A. Viewers will learn about drills.
   B. Viewers will buy a Sidewinder drill.

4. Name two common delivery systems for video programs:

_____

_____.

5. The organizing principle that shapes a program is called the _____.

6. Two of the three common levels of program treatment may be called

_____

_____.

T F 7. A storyboard may cover only certain parts of a program.

8. The following stages of script writing are out of order. To put them into correct order, write the letter of each stage in the appropriate blank.

_____ Stage 1   A. Revised second draft written and reviewed.

_____ Stage 2   B. Final polish of revised draft completed and approved.

_____ Stage 3   C. Agreement on concept and content.

_____ Stage 4   D. Detailed content outline written, reviewed, revised.

_____ Stage 5   E. First complete draft written and reviewed.

T F 9. Library footage shows reconstructions of events that were not originally shot.

10. Which off-camera voice does the term "interviewer" describe?
    A. A person reading scripted material.
    B. A person asking questions.

Activity 9-1

Developing a Program

Try developing a short video program by completing the steps discussed in the text. First, select a subject; then develop the concept that will guide your handling of that subject. Next, write a summary treatment of your program. Finally, draw a detailed storyboard for *one sequence* in your program.

Program Subject

Choose a subject for your program. You may select one of the listed suggestions or invent your own subject. Remember: a subject should be capable of being expressed as a *working title* for your program.

Suggested Program Topics

Stories

The Great Race

Buried Treasure

The Fateful Letter

Alien Attack

Documentaries

Sports Practice/Training (any sport)

The School Paper

Learning to Drive

A Complete Makeover

New Student at School

The Big Examination

School Election

Pick any of these subjects or invent your own, if you prefer. Whatever your subject, it should be substantial enough so that you can develop a concept, a treatment, and a storyboard.

Program Concept

Review Chapter 8 to refresh your memory about concepts. Then select a concept appropriate for your topic. Here is a sample.

Subject: School History.

Concept: McKinley High has grown with the town.

Before working on the topic you have selected or invented, practice developing concepts by choosing three subjects from the preceding list of suggestions. Write the subjects in the spaces below, then developing a concept for each.

Subject: _____

Concept:_____

Subject: _____

Concept:_____

Subject: _____

Concept:_____

Activity 9-2

Program Treatment

Write a summary treatment of the program based on the subject and concept you have developed. You do not have to describe every shot, but you should include a brief (usually one sentence) description of every major scene in the program. As an example, here is just the opening of the treatment for the McKinley High history program.

An opening montage of scenes around the present-day high school dissolves to still shots of old photographs showing its beginnings. Narration establishes the theme: how the high school grew with the town.

Historical footage of McKinley in the early 20th century: funny cars, funky clothes, wooden buildings, with ragtime music in the BG. Narration gives historical background about the early days of the town. Footage of teenagers establishes need for a high school. Shots of one-room schoolhouse show inadequacy of facilities.

Using the rest of this sheet, write a summary treatment of your program. Be sure to indicate how your concept will be expressed in the video. Continue on a separate sheet of paper, if needed.

_____

_____

_____

_____

_____

_____

_____

_____

_____

_____

_____

_____

_____

_____

_____

_____

_____

_____

_____

_____

_____

Activity 9-3

Storyboarding

Now pick a single sequence from your treatment for storyboarding. Make sure it is a fairly long sequence (using one frame for each shot, you should at least use up the 18 frames on this and the following page). To help explain your drawings, write brief notes underneath them, as shown in the example.

Charlie spreads the news that the school is burning down.

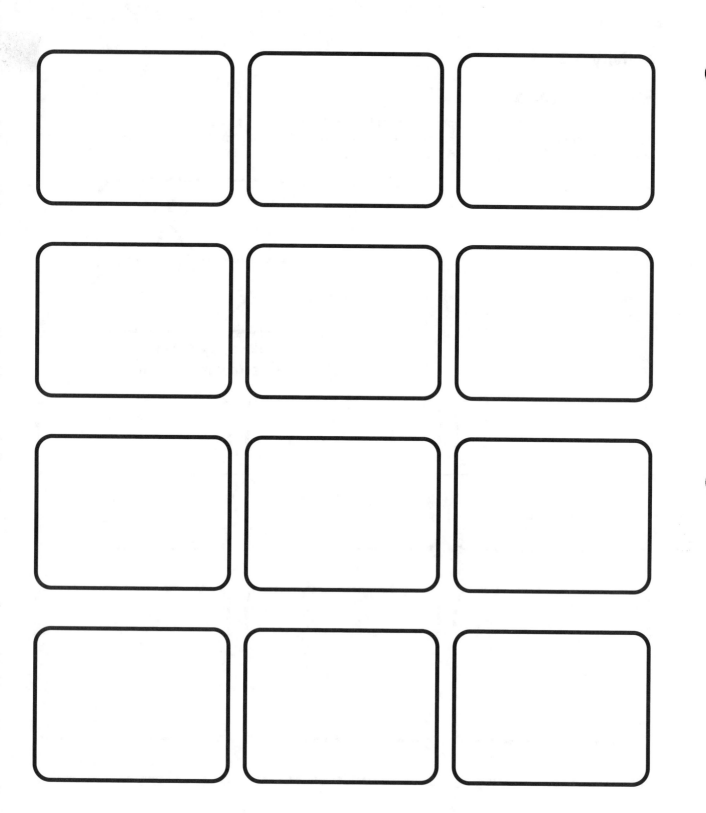

Activity 9-4

Mini-project: Hanging Wallpaper

Description

This activity shows how to create "wallpaper footage" to show while the content is carried by narration.

Objective

To provide practice in shooting material to fill screen time.

Materials

- Camcorder kit.
- Narration script.

Preproduction

1. Study the script below. Obtain the information needed to fill in the blanks and do so. (If necessary, you can change the script to fit your situation—or write your own script, *making sure that its content cannot be shown in live action.*)

2. Make a list of shots that could *plausibly* use screen time while the narration is playing. If you are equipped to record them, you can include stills, as appropriate.

Production

1. Videotape the narrator reading the narration. Place the camcorder for optimal sound quality, since you will not use the video. *Tip:* Remember that narration is read somewhat more slowly than conversational speech. Also, allow pauses after each sentence, to aid in timing during editing.

Note: This project switches back and forth between production and postproduction.

2. Edit the narration, adjusting timing and eliminating any mistakes.

3. Record the suggested visuals.

Postproduction

Edit the video footage, trimming and timing the shots to follow the narration.

Discussion

1. Where do you often see wallpaper footage on TV? Why?

2. Some shots seem more appropriate to narration than others. Why?

3. Could this topic be recast with interview answers instead of narration? How?

Script

_____ (name of college, school, other organization) _____ was founded in _____ _____ (year) _____ by _____ (group, govt. or organization) _____. Its present facilities were completed by _____ (year) _____. At first, the student body numbered around _____. (*number of students*)_____, and today, _____ (describe change – or no change– in number of students) The chief administrator is the _____ (title, such as President, rather than name of incumbent) _____, and other top-level administrative posts include _____ (job title, job, title, job title, and job title) _____. The academic year is divided into _____ (semesters, quarters, sessions) _____, each of which runs _____ (number of weeks) _____. There are _____ (number of) _____ these per year.

10 Production Planning

Reading Review

Write a brief answer for each question below. If you are not sure of an answer, review the appropriate section of the chapter to find it.

1. Explain why production planning is so important.

   _____

2. Identify three good sources for amateur crew members.

   _____

3. Name two ways to help conduct effective talent auditions.

   _____

   _____

4. Name three important things to look for in scouting locations.

   _____

   _____

   _____

5. List two important things to do in planning production equipment needs.

   _____

   _____

6. Describe three ways to obtain loyalty from volunteer cast and crew members.

   _____

   _____

   _____

7. What kinds of videos require talent releases and location permissions?

   _____

8. Explain why scheduling is so important to a smooth shoot.

   _____

9. List, in order of importance (most important first), the four categories by which you organize the shooting schedule.

   _____

   _____

10. Briefly explain why production budgets are important.

    _____

11. Explain what a contingency fund is.

    _____

Vocabulary Review

Match each term to its definition by writing the letter of the correct definition in the space provided. (Not every one of these terms is defined in the *Technical Terms* section in the textbook, and some of the definitions do not apply to any term listed here.)

Terms

_____ 1. Time code

_____ 2. Crew

_____ 3. Properties

_____ 4. Audition

_____ 5. Director

_____ 6. Budgeting

_____ 7. Cast

_____ 8. Videographer

_____ 9. Scouting

_____ 10. To dress

Definitions

A. The crew member responsible for lighting and taping the program.

B. Predicting the costs of every aspect of a production and allocating funds to cover it.

C. Every production member who performs for the camera.

D. Production staff members who work behind the camera.

E. The recorded numbers that identify each individual frame of video.

F. A session in which a potential cast member is evaluated for suitability.

G. The crew member who determines camera setups and shapes the performances of cast members.

H. Documents allowing the recording of people and private places.

I. An artificial environment built for use in shooting the program.

J. A "real world" shooting environment.

K. To add decorative items to a set or location. Such items are called "set dressing."

L. The people who appear in the program.

M. Things used on-screen in a video program.

N. Prerecording an entire tape with a pure-black picture and no sound, and time code (or, in analog formats, a control track).

O. The process of locating and evaluating potential shooting locations.

Chapter Quiz

Answer each question below. For True/False questions, circle the correct answer. For other questions, write the answer in the space provided.

T F 1. Production planning means preplanning shots to be made during the shoot.

2. The two groups of people who work on a video are

_____

_____.

3. What are casting auditions for?

_____

4. List two tips for successful auditions.

_____

_____

5. Explain why you should rely on audition videotapes in selecting performers.

_____

6. List three common problems with volunteer cast and crew members.

_____

_____

_____

T F 7. Never tell any performers how important they are because they will always use this power to do whatever they want to.

8. List the four things to evaluate when you scout locations.

_____

_____

_____

_____

9. List three common location problems with regard to video. _____

_____

_____

10. Explain why wireless microphones should be tested at each location prior to production.

_____

T F 11. Props are used to brace the camera tripod on steep slopes.

Name_____

Date_____

Activity 10-1

The Literate Viewer: Analyzing Production Planning

To get an idea of how much production planning goes into a professional program, try the following experiment. Record a one-hour series program. Avoiding situation comedies (which use limited sets and casts), select a dramatic program — a medical, crime, legal, political, or other type featuring a larger cast and a variety of settings. Now analyze *only the first half-hour* of the program recording and fill in the information called for below.

Locales. Name each locale in which a scene takes place. Be specific—for instance, instead of "hospital," you might write, "emergency room," "nurses' station," "administrator's office," etc.

Scenes. Enter the locale name for each scene (reusing names when locales are repeated). An old scene ends and a new scene begins when the locale changes or the time of day or the date changes.

Characters. Identify each character. If you don't know the name, use a description like, "Old Doctor," "1st Nurse," and so on. Collectively label everyone who doesn't speak as "extras."

When you have completed your analysis, each scene should look like the example below.

Scene #	Locale Name	Characters in Scene
1	accident scene	mother, daughter, other driver, 1st paramedic, 2nd paramedic, detective. 1st policeman, 2nd policeman, witness, extras.

Scene #	Locale Name	Characters in Scene

Activity 10-2

Creating a Shooting Schedule, Part 1

Now you can develop a shooting schedule for the scenes you recorded. To start the process, recopy your entries from the previous two pages, grouping them first by whether they are interiors or exteriors (inside or outside) and then by locale, so that all the scenes in each place are together. (You can omit the characters in this version.)

Scene #	Locale Name	Interior/ Exterior	Scene #	Locale Name	Interior/ Exterior

Activity 10-3

Creating a Shooting Schedule, Part 2

So far, you have grouped scenes by interior/exterior and then, within each category, by locale. The final step is to reorder the scenes in each locale, so that different actors can be scheduled to start work at different hours. (Refer back to your original worksheet for the characters in each scene.) In these sample scenes, notice how the characters overlap.

Scene 11	**Scene 14**	**Scene 17**
Manny	Moe	Curly
Moe	Curly	Larry
Jack	Larry	Alice

Scene #	Locale	Characters	Scene #	Locale	Characters

Camera Systems

Reading Review

Write a brief answer for each question below. If you are not sure of an answer, review the appropriate section of the chapter to find it.

1. Why is a portable video recording system called a "camcorder?"

   _____

2. Why do three-chip cameras generally produce better pictures than one-chip models?

   _____

3. What is the difference between a "field" and a "frame?"

   _____

   _____

4. What are the three main types of camera support systems?

   _____

5. What is the difference between "panning" and "tilting?"

   _____

6. Name the seven common adjustments you can make on a tripod head.

   _____

   _____

7. What is the main advantage of a stabilizer system? What is the main disadvantage?

   _____

   _____

8. What is a pedestal dolly?

   _____

9. List the three principal types of camcorder batteries.

   _____

10. What is the minimum number of batteries required to cover a shooting session?

    _____

11. Describe the difference between a CRT built-in monitor and a CCD built-in monitor.

    _____

12. Why is an external reference monitor desirable?

    _____

    _____

13. When shooting, how do you prepare and check a recording tape before use?

_____

_____

14. Name the four principal modes (settings) for white balance.

_____

15. What are the color temperatures of halogen lights and direct sunshine?

_____

16. Describe the procedure for cleaning a lens.

_____

_____

17. Why is some kind of filter always desirable on the front of the lens?

_____

18. What are the two most often used automatic/manual camera controls?

_____

19. How does electronic lens stabilization work?

_____

_____

20. How does optical lens stabilization work?

_____

_____

Vocabulary Review

Match each term to its definition by writing the letter of the correct definition in the space provided. (Not every one of these terms is defined in the *Technical Terms* section in the textbook, and some of the definitions do not apply to any term listed here.)

Terms

_____ 1. Audio

_____ 2. Camcorder

_____ 3. CCD (Charge-coupled device)

_____ 4. Camera dolly

_____ 5. Color temperature

_____ 6. Diopter correction

_____ 7. Drag (pan or tilt)

_____ 8. Field

_____ 9. Filter

_____ 10. Image stabilization

_____ 11. Interlace

_____ 12. Mini-DV

_____ 13. Pedestal dolly

_____ 14. Quick-release mechanism

_____ 15. Reference monitor

_____ 16. Stabilizer

_____ 17. Three-chip camera

_____ 18. Tripod

_____ 19. Video

_____ 20. White balance

Definitions

A. The images captured and recorded by a camcorder.

B. A piece of glass placed (usually) in front of the lens to alter the quality of the image being recorded.

C. Half the information in an interlaced frame of video, consisting of either all the odd-numbered scan lines or all the even-numbered ones.

D. A camera mounting system that reduces image shake when camcorders are hand-held.

E. A connection that allows the camcorder to be quickly attached to and detached from the tripod.

F. A microscopic sensor that turns light into electric current.

G. The sounds captured and recorded by a camcorder.

H. A camcorder that splits incoming light into red, green, and blue primary colors and records each one on a separate CCD.

I. The camera setting selected to compensate for the color temperature of the light source illuminating the subject.

J. A video playback set calibrated for accuracy and used while shooting to evaluate the images being recorded.

K. A moving camera platform with the camera mounted on a central telescoping column.

L. The camcorder imaging chip that converts optical images into electronic signals.

M. An optical device that divides incoming light into red, green, and blue components.

N. A device that both captures and records moving images and sounds.

O. A three-legged camera support that permits leveling and turning the camera.

P. An electronic device that converts the analog signal from the CCD to a digital signal for recording.

Q. A rolling camera support.

R. Compensation to minimize the effects of camera shake.

S. A mechanical arrangement that allows the user to level a tripod head without adjusting the tripod legs.

T. A tripod control for adjusting its resistance to moving horizontally or vertically.

U. The most common recording method (and tape format) used in small digital camcorders.

V. The overall color cast of nominally "white" light, expressed in degrees on the Kelvin scale.

W. A three-armed accessory that turns a tripod into a dolly by adding wheels.

X. To create a frame of video image by successively displaying two fields of information.

Y. Provision for adjusting a camcorder viewfinder to match the eyesight of the user.

Chapter Quiz

Answer each question below. For True/False or Multiple Choice questions, circle the correct answer. When more than one answer seems reasonable, choose the best one. For other questions, write the answer in the space provided

1. Two terms for the chip that converts the image into an electrical signal are

   _____ or _____.

2. In the NTSC television system used in North America, each complete frame of picture is made of two fields combined by a process called

   _____.

3. A tripod has three legs instead of four, so that

   _____

   _____.

4. Name any three of the many controls on a typical tripod head.

   _____

T F 5. Camera dollies can be operated only on special tracks.

6. Name any one of the several types of camera batteries.

   _____

7. List two of the advantages of an external viewfinder.

   _____

   _____

8. Identify one of the reasons why a reference monitor is useful.

   _____

T F 9. Outdoor light tends to be slightly reddish because of the sun.

10. Explain how to set white balance manually.

    _____

    _____

11. If possible, the front of a lens should always be covered by at least this filter:
    A. neutral density
    B. polarizer
    C. UV/1A/skylight

12. When first turned on, most camcorders set focus and exposure controls to _____.
    A. automatic
    B. manual

13. Two different systems for lens stabilization are _____ and

    _____.

14. Which item is not appropriate for cleaning a lens?
    A. lens brush
    B. clean facial tissue
    C. microfiber cloth
    D. photo lens tissue

Name_____

Date_____

Activity 11-1

Mini-project: Camcorder Controls

Description

This activity helps you learn how to find and operate key controls on your camcorder.

Objective

To successfully operate the camcorder white balance, exposure, and shutter controls.

Materials

- A camcorder kit.
- Camcorder instruction manual.

Preproduction

Some camcorder controls are physical switches, buttons, or knobs. Others are line items in menus that can be displayed on the viewfinder. It is very important to locate and master these controls before making shots with the camcorder.

For this reason, it is essential to get a copy of the camcorder instruction manual. If none is available, try finding a downloadable copy on the manufacturer's website. If that does not work, use systematic detective work to discover (and write down) the various controls.

Production

For both safety and convenience, place the camcorder on a tripod.

White balance controls

1. Identify the location on the viewfinder where white balance information is displayed.

2. Determine whether your camcorder defaults to auto white balance when turned on. Change the setting to manual. Turn the camcorder off and then on again. Does the white balance stay at the manual setting or revert to auto?

3. Practice changing the white balance from auto to indoor, outdoor, and fluorescent. Note the viewfinder words or icons used to indicate each setting. Consult the instruction manual for additional information about each setting.

Exposure controls

1. Identify the location on the viewfinder where exposure information is displayed. Does your camcorder display exposures in "f-stops" (like 2.8, 4, 5.6, 8, etc.)?

2. Find the exposure programs (typically on a screen menu). Practice setting exposure for "backlight."

3. Go through the other exposure programs. Note the screen label or icon used to identify the various programs. Consult the instruction manual to learn what each program does.

Shutter speed controls

1. Identify the location on the viewfinder where shutter speed information is displayed.

2. Find the shutter programs (typically on a screen menu). Practice setting shutter speeds manually. Note the screen label or icon used to identify them.

3. Go through the various shutter programs. Consult the instruction manual to learn what they do.

Postproduction

No postproduction is required.

Discussion

1. What kinds of programs generally require using the automatic white balance and exposure settings? Why?

2. Why does so much information appear in the viewfinder (and/or external view screen)? Why is it important?

3. What other major control system has not been covered by this activity? How could you find out about it?

Activity 11-2

Mini-project: Controlling White Balance

Description

This activity lets you test your camcorder's white balance controls in a variety of conditions.

Objective

To determine how well the "auto," "switched," and "manual" white balance settings render natural looking colors.

Materials

- A camcorder kit.
- Subjects, as needed.

Preproduction

Identify three locations:

1. An interior lit by incandescent lights (preferably studio movie lights).
2. An interior lit by overhead fluorescent lights (most classrooms will work, if you stay away from windows).
3. An exterior lit by daylight (direct sunlight is best).

Enlist 1–3 subjects for testing flesh tones (a variety of ethnicities is preferable, if possible).

Production

In each location, set up a single, tripod-mounted medium shot (waist-up) of your subject or subjects. Then, record a short take (roughly 30-seconds) of every shot listed below. Letter and record an identifying slate for each shot. Remove the slate after about five seconds. For example, the slate for one of the number 1 shots below would read "auto/outdoors."

1. Tape the shot with the white balance set to "auto."
2. Tape the shot with the white balance switch set to match the location: "indoor," (incandescent), "outdoor," (daylight), and "fluorescent" white balance control.
3. Tape the shot with the white balance set manually, as described in the textbook.
4. Tape the shot with the white balance set to one incorrect setting. For example, if you are outdoors, tape with set the white balance to "fluorescent."
5. Tape the shot with the white balance set to the other incorrect setting. For example, if you are outdoors, tape set the white balance to "indoor."

When you have finished, you should have taped the following shots in every location:

- Auto white balance.
- Switch-selected correct white balance.
- Manually set white balance.
- First switch-selected incorrect white balance.
- Second switch-selected incorrect white balance.

Postproduction

Review your footage, noting the slate for each white balance setting.

Discussion

1. Which method produced the most accurate white balance: auto, switched, or manual?

2. Using the switched method, which setting handled fluorescent lights better: fluorescent or outdoor?

3. Why might you purposely use the indoor setting outdoors? To find out, see *Shooting Day for Night* in Chapter 13 of the text.

Activity 11-3

Mini-project: Controlling Exposure

Description

Camcorders record light. This activity provides practice in controlling that light by adjusting exposure.

Objective

To explore the differences between automatic and manual exposure control.

Materials

- A subject.
- A camcorder kit.

Preproduction

Identify three shooting locations:

1. An interior with direct access out a door to the outside.
2. An exterior with a bright sky or light-colored building.
3. A nearby exterior in the same light, but with a dark background (dark buildings or greenery work well).

Production

Remember to slate each shot for later identification.

Working with auto exposure (interior/exterior location)

1. Set the exposure control to auto (most camcorders default to this setting).
2. With the camcorder on a tripod inside the interior, tape this action:
   the subject starts in the interior, moves to the door, opens it, and walks out into the exterior.
3. Hand-holding the camcorder, repeat the same subject action, but follow the subject through the door to the outside.

Working with locked manual exposure

1. With camcorder and subject both outside, allow the auto exposure control to set exposure. Then, lock that exposure by changing the camcorder setting from auto to manual.
2. Repeat shots 2 and 3, as described above.

Working with backlit subjects (exterior locations with bright and dark backgrounds)

1. With the subject standing in front of a bright background (sky or a light building with direct sun on it) and the exposure control set to auto, frame and record a full-length shot.
2. Using the exposure program menu (or the external camera control) enable the "backlight" exposure program. Then repeat the same shot of the subject.
3. Using the auto exposure setting, zoom in until the subject's face fills the frame. When the auto exposure has set the exposure, change to manual to lock it. Zoom back out to a waist shot and make the same shot.
4. Moving to the third location nearby, position the subject in front of the dark background, set exposure to auto, frame and record a medium shot as before.

Postproduction

No postproduction is necessary. However, if you wish to make a demonstration program, you can edit out the slates, replace them with titles superimposed over the shots, and provide explanatory titles.

Discussion

1. What are the results of the different methods of handling varying exposures?

2. If you were to choose to lock exposure at the outside level, how would you record prior action inside?

3. Each method of compensating for backlight has drawbacks. What are they?

4. If you could not move the subject in front of a dark background, how could you achieve good exposure on both face and light background?

Activity 11-4

Mini-project: Controlling Focus

Description

This activity provides practice in creating clear, sharp images by controlling focus.

Objective

To explore the differences between automatic and manual focus control.

Materials

- A subject.
- A camcorder kit.
- Yardstick.

Preproduction

Select two shooting locations:

1. A place with a background area behind pieces of foreground (see Figure 11-18 in the textbook).
2. A studio or classroom with available movie lights.

Production

Remember to slate each shot for later identification.

Working with autofocus

1. Set the focus control to auto (most camcorders default to this setting).
2. With the camcorder on a tripod positioned as shown in **Figure 1**, tape a full shot (head to foot) of the subject walking across the background area, past one or more pieces of foreground area.

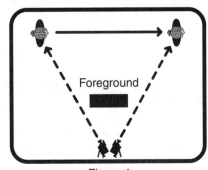

Foreground

Figure 1

Working with locked manual focus

1. With camcorder and subject both positioned as before, zoom in to fill the frame with the subject in the background area. Once the autofocus has set focus, switch the control to manual focus to lock it. Zoom back to the original composition.
2. Retake the shot, exactly as before.
3. Repeat both shots without the foreground objects. Instead, focus on subjects in the background, then have the subject walk through the shot close to the camcorder, as shown in **Figure 2**.

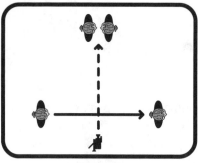

Figure 2

Working with depth of field

1. In studio lighting, set up the camcorder and yardstick as shown in **Figure 3**. Have a team member hold the yardstick, if necessary. Enable the aperture (f-stop) display in the camcorder finder.

2. Adjust the lighting until the aperture is somewhere between about f/1.8 and f/4, as displayed in the finder. You can do this either by adding or subtracting light units or by moving lights closer to or farther from the ruler.

3. Using the manual focus control, focus the camcorder at the 18-inch mark on the yardstick.

4. Record about two minutes of footage (so that the result is easy to study).

5. Adjust the lighting until the aperture is at least f/8 (preferably f/11 or f/16); then repeat the shot.

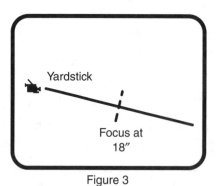

Figure 3

Discussion

1. In autofocus mode, what happens to the subject when a closer object appears in the foreground. How does the locking focus method correct this problem?

2. What is the sharp section of the yardstick called?

3. What is the difference in sharpness between the shot made at wide aperture (f/4 or wider) and the shot made at narrow aperture (f/8 or smaller)?

4. At each aperture, the yardstick appears sharp for a certain distance in front of the focus point (the 18-inch mark) and behind it. Are those distances the same? Which is greater?

Study the far-left diagram in the *Depth of Field* sidebar in Chapter 11 of the textbook. In general, the range of focus behind the plane of focus is twice as deep as the range of focus in front of it.

Camera Operation

Reading Review

Write a brief answer for each question below. If you are not sure of an answer, review the appropriate section of the chapter to find it.

1. Define the term "lens focal length."

   _____

2. What does "angle of view" mean?

   _____

3. How does a zoom lens differ from a "fixed focal length" lens?

   _____

4. Explain what determines whether a lens is wide angle, normal, or telephoto?

   _____

5. Explain how the maximum aperture of a lens is related to its "speed."

   _____

6. Calculate the aperture (f-stop) for a 40mm (millimeter) lens with a 5mm opening.

   _____

7. Name the three main optical characteristics of lenses.

   _____

   _____

8. Name the three main pictorial characteristics of lenses.

   _____

   _____

9. Explain the differences between wide angle and telephoto lenses in rendering apparent depth.

   _____

   _____

10. Give two reasons why lens autofocus can be unreliable.

    _____

    _____

11. Describe one way to focus a lens manually.

    _____

12. Explain why it is safer to focus at a lens' extreme telephoto position.

    _____

13. Why does the "gain" (or "gain up") setting let you shoot in lower light levels?

    _____

14. What is "camera drag" on a tripod used for?

_____

15. Give one of the five tips for steady hand-held shots.

_____

Vocabulary Review

Match each term to its definition by writing the letter of the correct definition in the space provided. (Not every one of these terms is defined in the *Technical Terms* section in the textbook, and some of the definitions do not apply to any term listed here.)

Terms

_____ 1. Polarizer

_____ 2. Autofocus

_____ 3. Specular reflections

_____ 4. Blocking

_____ 5. Neutral density filter

_____ 6. Optical zoom

_____ 7. Depth of field

_____ 8. Setting focus

_____ 9. f-stop

_____ 10. Perspective

_____ 11. Angle of view

_____ 12. Gain

_____ 13. Iris diaphragm (iris)

_____ 14. Magnification

_____ 15. Digital zoom

_____ 16. Speed

_____ 17. Aperture

_____ 18. Pulling focus

_____ 19. Shutter

_____ 20. Focal length

Definitions

A. The light-gathering ability of a lens, expressed as its maximum aperture.

B. The description of a zoom lens setting at any angle of view, expressed in millimeters (such as "25mm").

C. The apparent increase or decrease in subject size in an image, compared to the same subject as seen by the human eye.

D. The opening in the lens that admits light.

E. The name of a lens that, unlike a zoom, cannot change focal length.

F. The representation of depth in a two-dimensional medium.

G. The electronic amplification of the signal made from an image, in order to increase its brightness.

H. Changing the lens focus during a shot to keep a moving subject sharp.

I. Adjusting the lens to make the subject appear clear and sharp.

J. A lens whose images resemble those made by human vision.

K. Hard, bright reflections from surfaces such as water, glass, metal, and automobile paint.

L. A gray lens filter use to reduce excessive incoming light.

M. The camcorder system that automatically adjusts the lens to keep subjects sharp.

N. A particular aperture size.

O. Increasing the subject size by filling the frame with only the central part of the image.

P. The width of an image compared to its height.

Q. Changing the angle of view of a lens by moving internal parts of the lens.

R. The circuitry that determines how long the CCD will be exposed to light before an image is processed.

S. The characteristics of different lens focal lengths, as perceived by viewers.

T. A filter that can reduce reflections and darken blue skies.

U. A mechanism inside a lens that varies the size of the lens aperture.

V. The breadth of a lens's field of coverage.

W. The distance range, near-to-far, within which subjects appear sharp in the image.

X. A pattern of diagonal stripes indicating an overexposed part of the image.

Y. The predetermined movements that a subject and/or camera make during a shot.

Chapter Quiz

Answer each question below. For True/False questions, circle the correct answer. For other questions, write the answer in the space provided

T   F   1. The focal length of a lens determines whether it is wide angle, normal, or telephoto.

2. In determining the focal length of a lens, the front, middle, and rear points are called the

1) _____ , 2) _____ and 3) _____ .

T   F   3. Depth of field is the distance in front of the plane on which the lens is focused.

4. Explain the difference between a lens' *optical* and *pictorial* qualities.

_____

_____

5. An aperture is indicated by the letter _____, which stands for _____ .

6. Why does "digital zoom" produce an inferior quality image?

_____

T   F   7. A polarizer is a filter used *primarily* to reduce incoming light.

8. Switching from a telephoto to a wide angle lens will not increase depth of field for a particular composition. Why?

_____

9. The "normal" shutter speed for NTSC video is _____ .

10. Two examples of shutter programs are

_____

_____ .

T   F   11. An "ND3" neutral density filter reduces the aperture by 3 f-stops.

12. Reducing the light forces the lens aperture to open wider. Give two reasons for doing this.

_____

_____

13. Explain "maintaining lead room."

_____

14. List two techniques for smooth panning and tilting on a tripod.

_____

_____

Activity 12-1

Mini-project: Wide Angle Lens Settings

Description

This activity demonstrates the pictorial characteristics of wide angle lens settings.

Objective

To show how wide angle settings render depth, movement, and human features.

Materials

- Two subjects.
- A camcorder kit.

Preproduction

Identify an exterior location permitting a very long distance from the camera to the extreme background. An athletic field or a large parking lot will work well.

Production

Set up the camera at one end of the location with a clear view of its length. Set the lens to extreme wide angle. Position the subjects as shown in **Figure 1**.

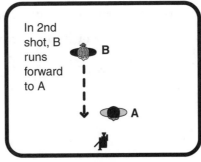

Figure 1

1. Tape several seconds of the two subjects standing in position.

2. Repeat the shot, but this time, have subject B run forward as fast as possible to join subject A. Continue the shot for a few seconds as they stand side by side.

3. Move subjects A and B back until you can frame them in a medium long shot, as in **Figure 2**. Have A and B take positions as shown in **Figure 3**. Record a shot, panning with A and B as they run at right angles to the camcorder.

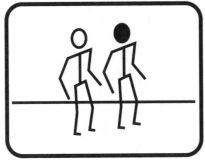

Figure 2

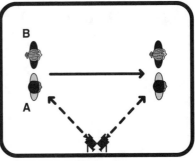

Figure 3

4. Move subject A forward until the image is a big closeup in 3/4 view, as shown in **Figure 4**. Record several seconds of this shot.

Figure 4

5. Place the camcorder at ground level, as close as possible to subject B's feet, with the external view screen rotated so that it is visible. Frame a worm's-eye angle and shoot several seconds.

Postproduction

Create clips of each shot and edit them in sequence for display. *Note:* Save these clips for re-use later.

Discussion

1. How deep does the location area appear on screen, compared to its actual length?
2. How quickly does subject B appear to run from starting to ending positions?
3. Describe the appearance of subject A's face in big closeup. What accounts for it?
4. Describe the appearance of subject B's face and body in the worm's-eye angle. What accounts for it?

Activity 12-2

Mini-project: Telephoto Lens Settings

Description

This activity demonstrates the pictorial characteristics of telephoto lens settings. Its procedures repeat those of the previous activity.

Objective

To show how telephoto settings render depth, movement, and human features.

Materials

- Two subjects.
- A camcorder kit.

Preproduction

Identify an exterior location permitting a very long distance from the camera to the extreme background. An athletic field or a large parking lot works well.

Production

Set up the camera at one end of the location with a clear view of its length. Set the lens to extreme wide angle. Position the subjects as shown in **Figure 1**.

(Camera and subjects are positioned as in wide angle activity.)

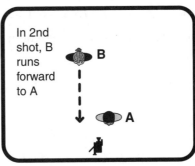

Figure 1

1. Tape several seconds of the two subjects standing in position.

2. Repeat the shot, but this time, have subject B run forward as fast as possible to join subject A. Continue the shot for a few seconds as they stand side by side.

3. Move subjects A and B back until you can frame them in a medium long shot, as in **Figure 2**. Have A and B take positions as shown in **Figure 3**. Record a shot, panning with A and B as they run at right angles to the camcorder.

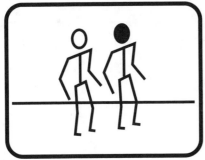

Figure 2

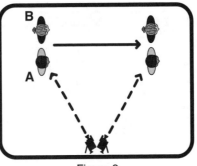

Figure 3

(Note that subjects must move much farther from the camera to frame a medium long shot at a telephoto lens setting.)

4. Move subject A forward until the image is a big closeup in 3/4 view, as shown in **Figure 4**. Record several seconds of this shot.

Figure 4

5. Place the camcorder at ground level, as close as possible to subject B's feet, with the external finder rotated so that it is visible. Frame a worm's-eye angle and shoot several seconds.

Postproduction

Create clips of each shot and edit them in sequence for display. *Note:* Save these clips for re-use later.

Discussion

1. How deep does the location area appear on screen, compared to its actual length?

2. How quickly does subject B appear to run from starting to ending positions?

3. Describe the appearance of subject A's face in big closeup. What accounts for it?

4. Describe the appearance of subject B's face and body in the worm's-eye angle. What accounts for it?

Activity 12-3

Mini-project: Comparing Wide Angle and Telephoto Lens Settings

Description

This activity re-edits clips created in the two previous activities, for shot-to-shot comparison.

Objective

To display the differences between wide angle and telephoto lens settings.

Materials

- The individual shot clips created previously for the wide angle and telephoto activities.

Preproduction

None

Production

None

Postproduction

Re-edit all the clips into a single program, pairing each wide angle shot with the same shot made in telephoto mode:

1. Wide angle/telephoto shots of A and B standing still, then B running to join A.
2. Wide angle/telephoto shots of the big closeup.
3. Wide angle/telephoto shots of the worm's-eye angle.
4. Wide angle/telephoto shots of A and B running at right angles to the lens.

Discussion

1. Considering the first pair of shots, which lens settings are generally preferred for fights, chases, and other dramatic movement? Why?
2. Considering the second and third pairs of shots, which lens settings are generally preferable? What other options do you have in setting up these shots?
3. Considering the last pair of shots, how much difference is there between the wide angle and telephoto versions? Why? What does this suggest about staging (directing) action scenes?

Activity 12-4

The Literate Viewer: Wide Angle and Telephoto Shots

Description

This activity explores ways in which wide angle and telephoto lenses are used in TV commercials.

Objective

To identify and collect as many wide angle and telephoto shots as possible.

Materials

- TV.
- VCR.
- Videotape.

Preproduction

None

Production

Tape at least two hours of prime-time (evening) TV. Network channels are generally good. Avoid specialty channels addressed to children, shoppers, etc. Programs with many automobile commercials are particularly good.

Postproduction

Review your tape. As you identify each wide angle or telephoto shot, make a clip of it. Then edit all the wide angle shots together, followed by all the telephoto shots.

Discussion

Screen your edited shots for other members of the class. Do they think you misidentified any shots? (Confusing wide angle and telephoto shots is surprisingly easy to do.)

1. Why do car commercials, especially, use wide angle shots?

2. Sometimes car commercials use extreme telephoto shots. Why do you think they do this?

Activity 12-5

Mini-project: Wide Angle and Telephoto Size and Distance

Description

This activity compares the effects of wide angle and telephoto lenses on subject size and apparent distance. (*Note:* This activity creates the shots shown in Figure 12-8 in the textbook.)

Objective

To control size and depth through lens selection and camera placement.

Materials

- Two subjects.
- A camcorder kit.

Preproduction

Identify an exterior location permitting a very long distance from the camera to the extreme background. An athletic field or a large parking lot works well.

Production

1. Position the camcorder and subjects A and B as shown in **Figure 1**. Note that the camcorder is at the midpoint of the location. Set the lens at extreme wide angle and frame a full shot of A with B in the background, **Figure 2**. Shoot several seconds of footage. *Hint:* Place B just to one side of A, so that B will appear in the telephoto version of this shot, as described below.

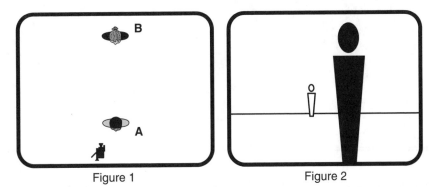

Figure 1   Figure 2

2. With subjects A and B standing in exactly the same spots, set the lens at extreme telephoto. Using the external view screen as a guide, move the camcorder back until A is framed *exactly* as in the previous wide angle shot and B is still visible behind him (**Figures 3 and 4**). Shoot several seconds of footage.

(Camera must be placed farther back than the diagram indicates.)

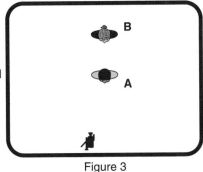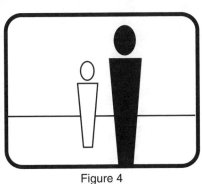

Figure 3   Figure 4

Postproduction

Create clips of the two shots and edit them together.

Discussion

1. Do subjects A and B seem closer together in the telephoto shot? Why?

2. If you were shooting a melodrama heroine tied to a railroad track with a train bearing down on her, which lens setting would you use? Why? How would you set up the shot?

3. Suppose you wanted to show two cars hurtling toward an inescapable head-on collision. Of course, you wanted the cars to actually pass each other with a safe distance between them. How would you stage the shot? When would you cut away from the shot? How could you simulate a terrible crash? (*Hint:* Sound is invaluable for this kind of special effect.)

Bonus Activity

If you can move the camera very smoothly (say, on a dolly) try this famous shot invented by Alfred Hitchcock:

With a telephoto lens setting, frame a full shot of a subject. Then record the shot, *dollying forward* while zooming *out* toward wide angle, *so that the subject remains the same size throughout the shot.* (You may need manual zoom control to make this work.) Then try the same trick in reverse, dollying in while zooming out. When you screen the results, what happens to the background? How might you use this trick in a video?

Lighting Tools

Reading Review

Write a brief answer for each question below. If you are not sure of an answer, review the appropriate section of the chapter to find it.

1. Define "video lighting."

_____

2. What three aspects of available light can you control?

_____

3. What are the five common types of lights used in video?

_____

_____

4. Explain the most important quality of spotlights.

_____

5. Explain the most important quality of softlights.

_____

6. What is a "practical"?

_____

7. Explain the difference between incandescent and halogen lamps.

_____

_____

8. What is a C-stand used for?

_____

9. Why is gaffer tape better than duct tape in video lighting applications?

_____

_____

10. Why is a reference monitor important in lighting design?

_____

Vocabulary Review

Match each term to its definition by writing the letter of the correct definition in the space provided. (Not every one of these terms is defined in the *Technical Terms* section in the textbook, and some of the definitions do not apply to any term listed here.)

Terms

_____ 1. Available light

_____ 2. Diffusion

_____ 3. Lamp

_____ 4. Screen

_____ 5. Spotlight

_____ 6. Barn doors

_____ 7. Instrument

_____ 8. Contrast

_____ 9. Broad

_____ 10. Flag

_____ 11. Key light

_____ 12. Kelvin

_____ 13. LED

_____ 14. Silk

_____ 15. Century stand

_____ 16. Halogen lamp

_____ 17. Cookie

_____ 18. Neutral density filter

_____ 19. Practical

_____ 20. Pan

Definitions

A. The natural and/or artificial light that already exists at a location.

B. A unit of lighting hardware such as a spotlight or floodlight.

C. A lamp with a filament and halogen gas enclosed in an envelope of transparent quartz.

D. The difference between the lightest and darkest parts of an image, expressed as a ratio (e.g., "four-to-one").

E. A small light mounted on the camera to provide foreground fill.

F. Light Emitting Diode: an electronic light source for special applications.

G. White spun glass or plastic sheeting placed in the light path to soften and disperse it.

H. Metal flaps in sets of two or four, attached to the fronts of spotlights to control the edges of the beam.

I. A flat piece of opaque metal, wood, or foam board placed to mask off part of a light beam.

J. A telescoping floor stand fitted with a clamp and usually an adjustable arm, for supporting lights and accessories.

K. A very large, flat light source, usually fitted with several fluorescent lamps.

L. A lamp with a filament enclosed in a glass envelope in a near-vacuum ("bulb").

M. The actual bulb in a lighting instrument.

N. A lamp or small light enclosed in a large fabric box, which greatly diffuses the light.

O. The scale used in measuring color temperature.

P. In lighting, a gray sheet filter placed over a window to reduce the intensity of the light coming through it.

Q. A light meter that measures illumination as it bounces off the subjects and into the camera lens.

R. A mesh material that reduces light intensity without markedly changing its character.

S. A small rectangular floodlight.

T. A fabric material that reduces both light intensity and directionality, producing a soft, directionless illumination.

U. A sheet cut into a specific pattern and placed in the beam of a light to throw distinctive shadows such as leaves or blinds.

V. A type of floodlight used mainly in TV studios.

W. A small-source lighting instrument that produces a narrow, hard-edged light pattern.

X. The principal light on a subject.

Y. An instrument that is included in shots and may be operated by the actors.

Chapter Quiz

Answer each question below. For True/False questions, circle the correct answer. For other questions, write the answer in the space provided

1. Three tools for controlling available light quantity are

   _____

   _____ .

T F 2. Window filters convert indoor color temperature to outdoor.

3. Two types of softlights are

   _____

   _____ .

4. The four types of lamps are

   _____

   _____ .

5. Match the light to its definition:

   _____ 1. Spotlight
   _____ 2. Floodlight
   _____ 3. Softlight

   A. A small light mounted on the camcorder.
   B. A large light.
   C. A light with a focusable lamp.
   D. A very large, diffuse light source.
   E. A light with a bowl-shaped reflector.

T F 6. The light in an umbrella is often aimed away from the subject.

7. Match the light control accessory to its definition:

   _____ 1. Barn doors
   _____ 2. Flag
   _____ 3. Diffusion

   A. Sheet material that softens the light.
   B. Sheet material that changes light color.
   C. Flaps mounted on a spotlight.
   D. An opaque card that blocks light.

8. Foam board is used for _____.
   A. reflectors
   B. flags
   C. both
   D. neither

9. Two lighting instruments that can be bought at hardware stores are

   _____

   _____ .

Activity 13-1

Mini-project: Building a Reflector

Description

This activity covers building three kinds of reflectors, step-by-step.

Objective

To show how to make permanent and temporary reflectors.

Materials

- Two pieces of one-inch thick foam board, each 24 × 36 inches (or as close to that as you can find). One piece white on both sides, the other piece white on one side and black on the other.
  Tip: Activity 13-3 also requires foam board sheets. Read it over now so that you can obtain all the material at the same time.

- Roll of heaviest-weight, widest-width aluminum cooking foil.

- Adhesive, double-faced , or transparent tape, as available.

Preproduction

Obtain the needed materials.

Production

1. Attach aluminum foil *shiny side up* to one side of the white/white board, as shown in **Figure 1**. Be careful to affix it as smoothly as possible. Overlap a second sheet if one sheet is not wide enough to cover the board.

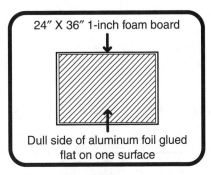

24″ X 36″ 1-inch foam board

Dull side of aluminum foil glued flat on one surface

Figure 1

2. Repeat the process on the other side of the board, but place the *dull side up* instead.

3. Cut a four-foot length of foil and *very gently* crumple it into a *very loose* ball. Working carefully, unfold the foil again and gently spread it as flat as possible. Do not attempt to smooth out the crinkles. (Repeat the process if more foil is needed.) Affix the foil, *dull side up* to the white side of the white/black board.

Discussion

1. How will the reflections of the smooth-shiny, smooth-dull, and crinkled-dull reflectors compare? Why?

2. Why was one board obtained with one black side, which was left bare? What could you do with a black card?

Activity 13-2

Mini-project: Building a Silk/Reflector Combination

Description

This activity will add a two-purpose tool to your lighting kit. *Note:* This activity requires access to a sewing machine.

Objective

To build a 4 × 6 foot frame and silk.

Materials

- Two 10-foot (standard length) pieces of 1 1/2 inch plastic plumbing pipe. (Several types are available. Test and then choose the most rigid.)
- Four 1 1/2 inch plastic pipe elbows.
- One discarded bed sheet. (Full size or larger.)
- A saw with a fine-toothed blade for cutting plastic.
- Sandpaper.

Tip: You can also make a screen for outdoor lighting that uses the same frame. Obtain black plant screening at a nursery or large home building supply store and prepare it like the sheet, as described below.

Preproduction

1. Obtain the materials.
2. Locate a sewing machine (and someone who can use it properly, if you cannot).

Production

1. Cut each 10-foot length of pipe to create a 4-foot piece and a 6-foot piece. Sand the rough edges of the cuts, especially on the outside surfaces.
2. Measure, cut, and hem the sheet as shown.

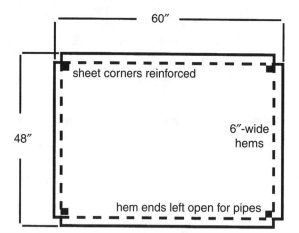

3. Assemble as follows:

- Place an elbow on one end of each pipe.

- Insert the other end of each pipe into the appropriate hemmed edge (4 or 6 feet long).

- Join all the pipes by inserting their free ends into the open sides of the elbows.

Discussion

1. Why is this tool so much larger than the foil reflectors?

2. How would you use it as a silk?

3. How would you use it as a reflector outdoors? Indoors?

Activity 13-3

Mini-project: Making a Flag or Cookie

Description

This activity constructs a basic tool for indoor shooting. *Note:* This activity requires the use of a sharp blade such as a utility knife.

Objective

To create one or more flags and/or cookies, as required.

Materials

- Several sheets of 24" × 36" foam board, white on one side, black on the other.
- Several 3/4" diameter wooden dowels, 24" long.
- Duct tape.
- Appropriate cutting blade.

Preproduction

Assemble materials for the project.

Production

1. Cut the foam board into several sizes, as shown in **Figure 1**.
2. If making a cookie, cut holes or patterns in the board as needed (**Figure 2**).
3. Tape the dowels to the boards, as shown in **Figure 3**.
4. Mount the flags/cookies on C-stands, when needed.

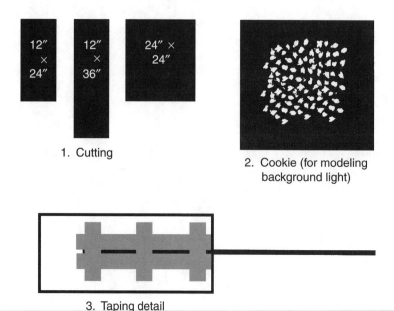

12"
×
24"

12"
×
36"

24" ×
24"

1. Cutting

2. Cookie (for modeling
background light)

3. Taping detail

Discussion

1. What are flags used for? What do cookies do?

2. Why make flags of different sizes?

3. Why is a wide border used around the cutout pattern in the cookie?

Activity 13-4

Mini-project: Testing Fluorescent Lighting

Description

This activity checks the color temperature of ceiling fluorescent lights in your working environment.

Objective

To obtain an appropriate white balance for videotaping.

Materials

- Camcorder outfit.
- Subject.

Preproduction

You complete many shooting projects in the environment where your courses are conducted. To see whether you can use the ambient light in your environment, you need to test several samples.

1. Identify several rooms or hallways lit with overhead fluorescent lights (probably including your classroom).
2. Make arrangements to shoot in each one.

Production

In each selected area, with all the fluorescent lights on, do the following (slate each shot with a scratch pad and felt marker):

1. Tape the subject in the environment with white balance set to "automatic."
2. Repeat with white balance set to "fluorescent."
3. Repeat with white balance set to "outdoor."
4. Set white balance manually:
   - Hold a white card in front of the camera at a 45° angle, so that it is lit by the ceiling lights.
   - Zoom as needed to fill the frame with the white card.
   - Set white balance to "manual."
5. Repeat the shot with the camcorder manually balanced for the room.

Postproduction

View and evaluate the results.

Discussion

1. Of the white balance presets, which seems to yield the most pleasing color balance?
2. How much variation is there in the color balance of lights in different locations?
3. How does the subject's face look when lit only by overhead fluorescents?
4. How could you improve the lighting?
5. How well would reflectors work? Why?

Lighting Design

Reading Review

Write a brief answer for each question below. If you are not sure of an answer, review the appropriate section of the chapter to find it.

1. Managing light quality means controlling three things. What are they?
   _____

2. What must a camera's imaging chip(s) receive in order to form a high-quality image?
   _____

3. What do you use outdoors to decrease light quantity? What do you use to increase it?
   _____
   _____

4. Define the term "contrast."
   _____

5. Explain two ways to control excessive contrast.
   _____
   _____

6. Give two reasons for controlling color.
   _____

7. Define mood and style in lighting.
   _____
   _____

8. Explain the difference between the lighting styles called "naturalism" and "realism."
   _____
   _____

9. Explain the different effects of screens vs. silks.
   _____

10. Describe small source lights.
    _____
    _____

11. Describe large source lights.
    _____
    _____

12. Explain the purpose of a Century stand.
    _____

13. Explain how to use a Color Rendering Index (CRI).

_____

_____

14. Explain why gaffer tape works better than ordinary duct tape.

_____

15. How do you use a reference monitor?

_____

Vocabulary Review

Match each term to its definition by writing the letter of the correct definition in the space provided. (Not every one of these terms is defined in the *Technical Terms* section in the textbook, and some of the definitions do not apply to any term listed here.)

Terms

_____ 1. Silk

_____ 2. Softbox

_____ 3. Barn doors

_____ 4. Spotlight

_____ 5. Cookie

_____ 6. Gels

_____ 7. Flag

_____ 8. Practical

_____ 9. Pictorial realism

_____ 10. Camera light

_____ 11. Reflector

_____ 12. Screen

_____ 13. Naturalism

_____ 14. Realism

_____ 15. Contrast

_____ 16. Incident meter

_____ 17. Floodlight

_____ 18. Century stand

_____ 19. Light meter

_____ 20. Lamp

Definitions

A. Metal flaps used to control the edges of spotlight beams.
B. A light mounted on the camera to provide foreground fill.
C. A telescoping floor stand for supporting lights and accessories.
D. A light meter that measures actual color of nominally "white" light.
E. The difference between the lightest and darkest parts of an image.
F. A patterned sheet placed in the beam of a light to throw distinctive shadows.
G. A flat piece of opaque material placed to mask off part of a light beam.
H. A light meter that measures the illumination as it comes from the light sources.
I. A unit of lighting hardware, such as a spotlight or floodlight.
J. The actual bulb in a lighting instrument.
K. An instrument that is included in shots and may be operated by the actors.
L. A light meter that measures illumination as it bounces off the subjects.
M. A silver, white, or colored surface used to bounce light onto a subject.
N. A mesh material that reduces light intensity.
O. A fabric material that produces a soft, directionless illumination.
P. A lighting style so lifelike that it's invisible to the viewer.
Q. A lighting style that heightens naturalistic lighting somewhat for dramatic effect.
R. A lighting style that "paints with light" for a more theatrical style of realism.
S. A style created by lighting and digital image modification to produce an unrealistic effect.
T. A lighting style designed to express powerful feelings, with little or no reference to realism.
U. Plastic filters used to color the light from windows or lighting instruments.
V. A small light source with a hard-edged, controllable output.
W. A large light source with a soft-edged, diffuse output.
X. A large light source created by placing a light inside a square cloth tent.
Y. An instrument for measuring the intensity of lights.

Chapter Quiz

Answer each question below. For True/False or Multiple Choice questions, circle the correct answer. When more than one answer seems reasonable, choose the best one. For other questions, write the answer in the space provided.

T   F   1. Video camcorders don't record images of people and places; they record light.

2. Lighting quality is controlled by managing three things:

_____.

3. List three ways to reduce excess contrast.

_____

_____

_____

T   F   4. Style is the emotional feeling of a scene.

5. Match the five main styles of lighting to their descriptions.

_____ Naturalism

_____ Realism

_____ Pictorial realism

_____ Magic realism

_____ Expressionism

A. A lighting style designed to express powerful feelings, with little or no reference to realism.

B. A style created by lighting and digital image modification to create an unworldly or otherwise unrealistic effect.

C. A lighting style so lifelike that it's invisible to the viewer. Useful for documentary and ultrarealistic effects.

D. A lighting style that "paints with light" for a more theatrical effect that is still based on natural light.

E. A lighting style that heightens naturalistic lighting somewhat for dramatic effect.

6. Explain the difference between a screen and a silk.

_____

T   F   7. Silver reflectors are suitable for lighting faces when placed far enough away from them.

8. Small-source lighting instruments are called _____.

9. What are "practicals?"

_____

10. What is the color temperature of halogen lamps?

_____

11. Spotlights may be fitted with _____.
   A. two barn doors
   B. four barn doors
   C. either two or four

T   F   12. A "cookie" is a flat panel used for masking off unwanted light.

Activity 14-1

Mini-project: Mastering Reflectors

Description

This activity uses reflectors outdoors for a variety of lighting jobs.

Objective

To use reflectors effectively outdoors.

Materials

Note: You need not make reflectors if they are already part of your equipment.

- Two pieces of white foam board, 24″ × 36″ or similar (24″ × 24″ minimum). Get the thickest foam board (up to one inch) available. White cardboard can be used but is harder to work with. You will mount aluminum foil to one side of both boards.
- Roll of aluminum cooking foil. The wide rolls intended for roasting, etc. are better, but you can use multiple strips of narrower foil if needed.
- Spray adhesive, double-faced tape, or transparent tape, as available.
- Subject.
- A camcorder kit.

Preproduction

1. Prepare your reflectors as described in Activity 13-1.

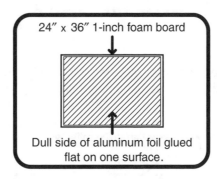

2. Choose a day with strong, direct sunshine (look for dark, hard-edged shadows).

Production

CAUTION!! ***Never*** aim a silver reflector at a subject's eyes from a close distance. To avoid eye discomfort or even injury, keep silver reflectors at least 15-20 feet from the subject when using them as key lights.

1. Position subject, sun, camcorder, and silver reflector so the reflector is the *key* light, **Figure 1**. Shoot several seconds of a medium closeup (head/shoulders/chest). Repeat this setup, using the white reflector instead of silver.

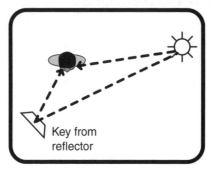

Figure 1

2. Position subject, sun, camcorder, and silver reflector so the reflector is the *fill* light, **Figure 2**. Shoot several seconds of a medium closeup (head/shoulders/chest). Repeat this setup, using the white reflector instead.

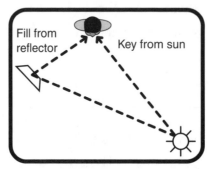

Figure 2

3. Position subject, sun, camcorder, and silver reflector so the reflector is the *back* light, **Figure 3**. Shoot several seconds of a medium closeup (head/shoulders/chest).

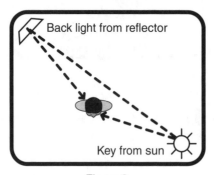

Figure 3

4. Position subject, sun, camcorder, and silver reflector so the reflector is the key light, **Figure 4**. Use the second silver reflector to splash a light pattern on the dark background. Shoot several seconds of a medium closeup (head/shoulders/chest).

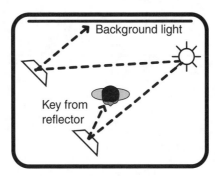

Figure 4

Postproduction

Make clean clips of your shots and assemble them for viewing.

Discussion

1. In which types of setup do reflectors seem to work best? Why?

2. How do your lighting techniques change when using white reflectors?

3. How would you handle sunlight and reflectors for the most "naturalistic" look? (Review the discussion of naturalistic lighting in the textbook.)

Activity 14-2

Mini-project: Classic Three-point Lighting

Description

This activity provides practice in the most common lighting setup for interviews, on-camera spokespersons, and other subjects who stay in one place.

Objective

To master the essentials of three-point lighting.

Materials

- Subject.
- A camcorder kit.
- A white foam board or cardboard reflector, preferably 36″ × 36″ or larger.

Suggested lighting instruments consist of:

- **Key light**: Spotlight with barn doors, or flood or soft light.
- **Fill light**: Flood or soft light, or spotlight used with spun glass or similar diffusion.
- **Back light**: Focusable spotlight with barn doors or small-source light masked off camera lens.

Preproduction

Assemble lights, light stands, diffusion, reflector.

Position the subject. *Hint:* A high stool is more comfortable for the subject than standing.

Production

These directions use a clock metaphor, with the subject at the center and the lights placed on the various numerals. Study **Figures 1** and **2,** which diagram this setup.

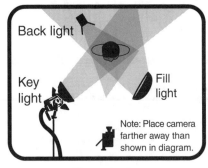

Figure 1: plan

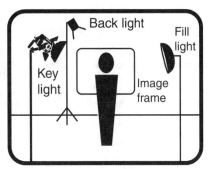

Figure 2: view

Place the subject on the stool about ten feet in front of a background wall. Set the camcorder and tripod as far away from the subject as practical and frame a medium closeup (head through chest).

Three-point lighting

1. Place the key light about 18″ higher than subject eye level, and anywhere from 7 to 9 o'clock (start at about 8 and then adjust to suit). Move the light closer to or farther from the subject until the key side of the subject's face is bright but not excessively so. (Check the camcorder viewfinder as you adjust the light.)

2. Place the fill light at eye level or a few inches above, and anywhere from 3 to 5 o'clock. Start with the light about 6′ from the subject, then move it closer to or farther from the subject until the fill side of the face is somewhat darker than the key side. Facial details should still be clearly visible. Note the effect (if any) of the fill light on the background behind the subject. Adjust the fill light to taste.

3. Place the back light behind the subject, as high as your equipment permits and as near the subject as possible without getting its stand or cable in the frame. Check the camcorder lens for flare from this light and shade the light or the lens. If you can adjust the light's distance and/or focus and/or voltage, adjust its brightness until the subject's hair and shoulders are "dusted" with light, but are not bright enough to look artificial.

4. Record 30 seconds of this setup.

Two-point lighting

For a softer, less dramatic effect, try two-point lighting, **Figure 3**, which is very common for interviews, and similar situations.

1. Turn off the back light.

2. Replace the key light spot with a flood or soft light, or place diffusion material in front of the spotlight.

3. Move the key light closer to the subject, to compensate for the less intense light. Then adjust the fill light, if necessary, to keep the fill side darker but fully detailed.

4. Record 30 seconds of this setup.

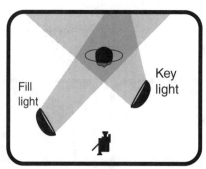

Figure 3

Two-point lighting with one light

This setup uses a reflector in place of the second light, **Figure 4**.

1. Set up the spotlight key, as in three-point lighting.

2. Replace the fill light with the large white reflector.

3. Adjust key light and reflector angles (around the "clock face") to maximize the bounce light from the reflector.

4. Adjust the distances of the key light and the reflector to maintain the same relative brightness and contrast ratio on the key and fill sides of the subject's face.

5. Record 30 seconds of this setup.

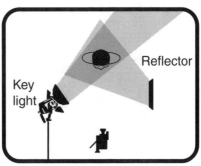

Figure 4

Postproduction

Create clean clips of about 10–15 seconds of each shot and assemble them for viewing.

Discussion

1. Which lighting setups produced the most "natural" look? Why?

2. Why is it better to position the camcorder fairly far away from the subject and lights?

3. How did the background appear in these setups? What are some ways to change the appearance of the background?

Lighting Applications

Reading Review

Write a brief answer for each question below. If you are not sure of an answer, review the appropriate section of the chapter to find it.

1. Explain the basic difference between studio lighting and natural lighting.

   _____

   _____

2. Explain how a clock face metaphor describes a lighting setup.

   _____

3. What does a fill light fill, and why?

   _____

   _____

4. Why is a rim light hung overhead, if possible?

   _____

   _____

5. How can you "help out" a natural, one-light design?

   _____

   _____

6. In lighting large areas, why is it often easier to light wide shots before closer ones?

   _____

   _____

   _____

7. Why is lighting wide shots easier outdoors and in sunshine?

   _____

8. Explain "back-cross" lighting.

   _____

   _____

   _____

9. In shooting day-for-night, why is it *not* necessary to set white balance for indoor if there is no artificial light on the scene?

   _____

   _____

   _____

10. Explain why "cheating" is a useful technique.

   _____

   _____

   _____

Vocabulary Review

Match each term to its definition by writing the letter of the correct definition in the space provided. (Not every one of these terms is defined in the *Technical Terms* section in the textbook, and some of the definitions do not apply to any term listed here.)

Terms

_____ 1. Amp

_____ 2. Day-for-night

_____ 3. Diffusion

_____ 4. Spill

_____ 5. Bounce

_____ 6. Fluorescent lamp

_____ 7. Glamorous lighting

_____ 8. Incandescent lamp

_____ 9. Magic hour

_____ 10. Neutral density filter

_____ 11. Rim light

_____ 12. Rugged lighting

_____ 13. Softlight

_____ 14. Tabletop

_____ 15. Tent

_____ 16. Three-point lighting

_____ 17. Throw

_____ 18. Voltage

_____ 19. Watt

_____ 20. Incidence

Definitions

A. A flat piece of opaque metal, wood, or foam core placed to mask off part of a light beam.

B. In lighting, the power rating of a lighting instrument 500, 750, and 1,000 watt lamps are common.

C. So-called "classic" subject lighting, consisting of key, fill, rim, and background lights.

D. The period, of up to two hours before sunset, characterized by long shadows, clear air, and warm light.

E. White spun glass or plastic sheeting placed in the light path to soften and disperse it.

F. A small light mounted on the camera to provide foreground fill.

G. The electrical potential or "pressure" in a system, nominally 110 volts in North America.

H. Lighting that emphasizes three dimensional qualities and surface characteristics of a subject.

I. Light falling onto an area where you do not want it.

J. A lamp with a filament enclosed, in a near-vacuum, in a glass envelope ("bulb").

K. A lamp or small light enclosed in a large fabric box, which greatly diffuses the light.

L. The distance between a light and the subject lit; also, the maximum useful distance between those points.

M. A white fabric draped all around a subject to diffuse lighting completely for a completely shadowless effect.

N. A lamp with a filament and halogen gas enclosed in an envelope of transparent quartz.

O. A light placed high and behind a subject to create a rim of light on head and shoulders, to help separate subject and background.

P. A method of shooting daylight footage so that it appears to have been taken at night.

Q. A lamp that emits light from the electrically charged gasses inside it.

R. Videography of small subjects and activities on a table or counter.

S. The angle at which light strikes a surface.

T. Lighting that emphasizes a subject's attractive aspects and de-emphasizes defects.

U. A small-source lighting instrument that produces a narrow, hard-edged light pattern.

V. The light that lightens shadows created by the main (key) light.

W. Light reflected off a wall, ceiling, or other surface.

X. In lighting, the amount of power being drawn by a light.

Y. In lighting, a gray sheet filter placed over windows to reduce the intensity of the light coming through them.

Chapter Quiz

Answer each question below. For True/False or Multiple Choice questions, circle the correct answer. When more than one answer seems reasonable, choose the best one. For other questions, write the answer in the space provided.

1. The four lights in a classic lighting setup are

_____

_____ .

2. The videographer can add to the effect of "rugged" lighting by using

_____ .

3. Name two of the things you are working to achieve in lighting backgrounds:

_____

_____ .

T F 4. A "hot" light beam means that it is putting a lot of heat on the subject.

5. Name three typical problems in lighting small interiors.

_____

_____

_____

6. In lighting close shots of dark-complected subjects in light clothing, you can
   A. mask the light off the clothing with a bottom barn door.
   B. place a half-screen in the lower part of a spotlight beam.
   C. either or both.

7. How can you make an off-screen "lighted sign" blink on and off?

_____

8. When lighting a day-for-night shot, the four steps are

_____

_____ .

T F 9. Lighting an interview with spots and broads is quicker because you have greater light control.

10. Give two reasons why fluorescent pan lights are useful in tabletop lighting.

_____

Activity 15-1

Mini-project: Lighting an Interview

Description

This project shows how to light two seated subjects with only two lights.

Objective

To experiment with several approaches to lighting an interview.

Materials

- Two soft lights. You can use (in order of preference) softboxes, umbrellas, broads or scoops with diffusion, or spotlights with diffusion. You can also use one type of instrument for the first light and another for the second light.
- Diffusion material. With spots, clip translucent plastic or spun glass sheets loosely to the barn doors. With scoops or broads, clip or hang diffusion material a safe distance in front of the lamps.
- Two subjects.
- A camcorder kit.

Preproduction

Create a simple set with a wall or curtain background and two subjects seated on chairs, "opened" toward the camera. Place lights on either side, ready for positioning. Place the camera to record a neutral two-shot (**Figures 1** and **2**).

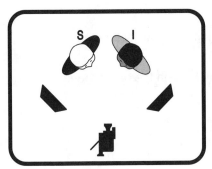
Figure 1

Figure 2

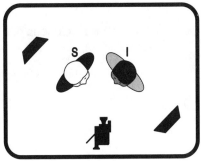
Figure 3

As you complete the exercise, you will move the camera to three additional setups. In the diagrams, "S" means subject and "I" means interviewer.

Production

Figures 1 and **3** suggest two different lighting designs. *Using each design in turn,* optimize the lighting for each of the four camera setups described. Record footage as the subject and the interviewer carry on a conversation. (Pause the conversation while changing camera setups.)

1. Light and shoot the neutral two-shot (Figures 1 and 2). Adjust the lights until the camera sides of the faces are brighter, but the off-camera sides are well lit, too.

2. Light and shoot a closeup of the subject (**Figure 4**). Adjust the lights for best results, while matching the overall look of the two-shot.

Figure 4

3. Light and shoot an "over-the-shoulder" two-shot of the subject (**Figure 5**). Place the camera so that the interviewer's mouth is off-camera. Adjust the right-side light so that the interviewer is rim-lit, but most of his/her back is darker.

Figure 5

4. Light and shoot a closeup of the interviewer (**Figure 6**). Again, match the overall look of the neutral two-shot.

Optional: Try adding lights to create backlight on the subject and modeling light on the background.

Figure 6

Postproduction

Edit the interview, alternating among the various shots.

Discussion

1. Which lighting scheme appears to give the best results? Why?

2. If you used one or more spots, broads, or scoops, how well did they work? What are the advantages of using large, soft light sources for simple lighting?

3. What are some reasons for shooting an over-the-shoulder two-shot of the subject? How could the editor use the fact that the interviewer's mouth is not visible?

Activity 15-2

Mini-project: Handling Mixed Light Sources

Description

This activity provides practice in lighting a location with lights of different color temperatures. *Note:* How you complete this activity will depend, not only on the kinds of movie lights you have, but also on the layout of the room location where you will shoot. For this reason, the following instructions are only approximate.

Objective

To solve the problems created by the different color temperatures of daylight, fluorescent light, and movie lights.

Materials

- Lighting instruments: Spotlights, soft lights, or umbrella lights (scoops and broads are difficult to use). Halogen work lights will also work.
- Reflectors: Large (up to 36″ square) hard silver and white. Inch-thick foam board works well, both as a white reflector and as a base for an aluminum foil reflector.
- Gels: Heat resistant blue transparent plastic sheets. Depending on the design of your lights, they can be mounted in rings that fit in front of a spotlight or clipped (using wooden clothespins) to the barn doors. Use them also on the spotlights inside soft lights or aimed into umbrella lights.
- A camcorder kit.
- Accessories.

Preproduction

Locate a room to use as a shooting location. It should have:
- A large window (ideally with movable draperies—either the kind that filter the light or the type that block it almost completely). A window without draperies is second-best. If you cannot locate a room with a large window, a small one is better than none at all.
- General lighting by fluorescent lights, either in hanging fixtures or above a ceiling grid.
- Enough space for the camcorder and lights to light a medium (waist) shot of a subject standing or seated at a desk.
- Enough electrical power for at least two, and preferably three, movie lights.

Figure 1 shows a representative location.

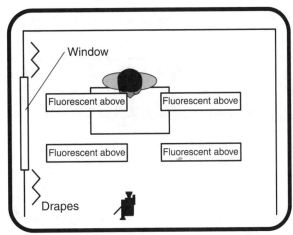

Figure 1

Production

You will light your subject with a classic three-light design, using movie lights covered by blue conversion gels plus the window and/or fluorescent illumination. Set white balance manually. (If your camcorder lacks this feature, set it to "outdoors.") The overhead fluorescents work best as fill light. As for the other light sources, you have several choices. You can:

- Key with the window light or with a spotlight.
- Turn off the fluorescent lights altogether and work with window and movie lights.
- Reduce or block the window light and work with movie lights and fluorescents.
- Eliminate both window and fluorescent lights and work with movie lights alone.

The following figures show lighting plans that you might consider adapting to the room you find and the movie lights at your disposal.

Figure 2 shows a simple approach: key the subject with light from the window and fill the dark side with a reflector. The overhead fluorescents provide general fill and lighting for the background. Experiment with both silver and white reflectors until the fill side is darker than the key side, but details are still clearly visible. (If direct sun is coming through the window, be very careful to avoid hurting the subject's eyes with the silver reflector.)

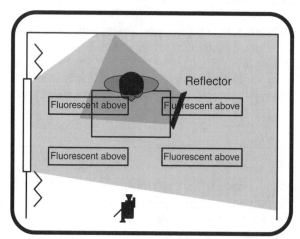

Figure 2

The **Figure 3** design turns off the fluorescents and adds a second reflector to throw a light pattern across the background. This approach provides a very "natural" look to the lighting.

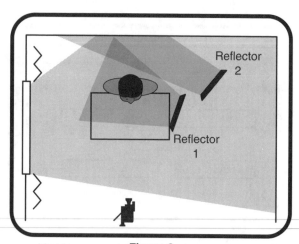

Figure 3

When you work with movie lights (**Figure 4**), try keying with a spotlight and using the window light as fill. Notice that a backlight (also called a *rim light*) separates subject from background. This design is useful when the overhead fluorescents must be turned off because they do not look good in an outdoor white balance setting.

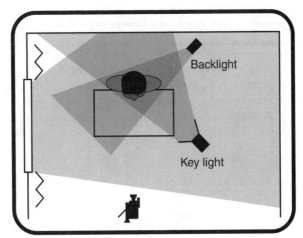

Figure 4

The **Figure 5** design also uses key and backlights; but it uses the fluorescents for extra fill, creating an overall "high key" look.

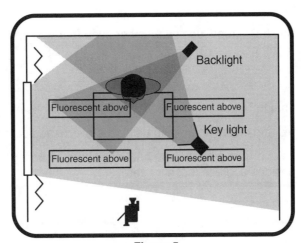

Figure 5

If the window light is inadequate and the fluorescents have poor color, the solution is to light entirely with movie lights (**Figure 6**). If the window cannot be covered by heavy draperies, try masking it with blankets or large sheets of foam board. In this case, blue gels on the movie lights are not needed. Set camcorder balance to match the movie lights without gels.

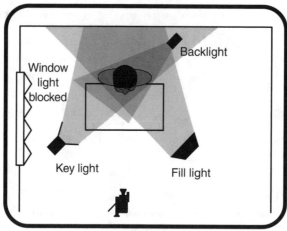

Figure 6

Postproduction

Assemble the footage recorded of the various lighting designs. If your editing software permits:

- Identify each design with a title.

- After displaying each design full-screen, use two-way or even four-way split screen to display the different lighting options together.

Discussion

1. Which design looks most natural (that is, "unlighted")? Which one looks most theatrical (studio-like)?

2. What are some advantages and disadvantages in using the overhead fluorescents?

3. What were some problems or other concerns you had in lighting the background with the lights available? How did your designs address these issues?

Activity 15-3

Mini-project: Lighting Day-for-Night

Description

This activity provides practice in creating lighting designs for day-for-night shooting.

Objective

To create an outdoor daytime design that looks like night when videotaped.

Materials

- A halogen light (spotlight, broad, or work light).
- A 110 volt ac power source reachable from outdoors.
- Silver reflectors.
- Subject and camcorder kit.

Preproduction

1. Scout to find a location with outdoor power, or else a ground-floor window through which a halogen lamp can be aimed from inside.

Production

1. Lock the camcorder white balance in the "outdoor" (daylight) setting.

2. Stage your action and set your camcorder for back cross light:
   - Use high, direct sunshine for rim light.
   - Use a reflector opposite the sun for fill light.
   - Make sure that the sky is not visible in the frame.

3. Set the exposure and lock it.

4. Add the halogen light, aimed from the side opposite the fill. Watching the monitor, move the light forward until it is distinctly brighter on the subject than the daylight (but not glaring).

5. Record footage for reference. *Tip:* It is often more effective to place the halogen light opposite a pausing point in the action. Start with the subject lit by daylight only and then move into the path of the halogen light, pause, and move forward again.

Postproduction

1. Assemble the footage. If needed, lower the exposure and increase the contrast slightly to enhance the moonlight look.

2. Review and discuss the results.

Discussion

1. How effective is the result? What (if anything) looks unrealistic about it?

2. Could you light day-for-night on an overcast day? Explain your answer.

3. Considering the fact that you are editing digitally, how would you change the setup if you did *not* have a halogen light in the scene? Why?

Recording Audio

Reading Review

Write a brief answer for each question below. If you are not sure of an answer, review the appropriate section of the chapter to find it.

1. Name the two biggest problems in recording audio.

_____

_____

2. List the three steps involved in addressing audio recording problems.

_____

_____

_____

3. Name any three of the five main types of audio recording.

_____

4. Match each microphone name with its description by writing its letter on the correct line.

_____ Condenser microphone   A. Rugged, simple, less accurate sound quality.

_____ Dynamic microphone   B. More accurate sound reproduction, delicate, needs a power supply.

5. List one drawback of cabled microphones.

_____

6. List one drawback of wireless microphones.

_____

7. Match each microphone type with its description by writing its letter on the correct line.

_____ Shotgun   A. Spherical sound pickup pattern.

_____ Lavaliere   B. Narrow, directional sound pickup pattern.

_____ Cardioid   C. Heart-shaped sound pickup pattern.

8. Explain what "level" means in sound recording.

_____

9. Explain why low-impedance microphones and cables are superior to high-impedance types.

_____

_____

10. Describe an easy way to tell unbalanced line cables from balanced line cables.

_____

11. List two tips for recording good quality room tone.

_____

_____

Vocabulary Review

Match each term to its definition by writing the letter of the correct definition in the space provided. (Not every one of these terms is defined in the *Technical Terms* section in the textbook, and some of the definitions do not apply to any term listed here.)

Terms

_____ 1. Transducer

_____ 2. Slate

_____ 3. Foley studio

_____ 4. Production sound

_____ 5. Cardioid

_____ 6. Playback

_____ 7. Fishpole

_____ 8. Wireless mike

_____ 9. Audio

_____ 10. Boom

_____ 11. Lavaliere

_____ 12. Microphone

_____ 13. Windscreen

_____ 14. Pickup pattern

_____ 15. Mixer

_____ 16. Unbalanced line

_____ 17. MOS

_____ 18. Equalizer

_____ 19. Balanced line

_____ 20. Looping

Definitions

A. A microphone cover designed to reduce wind noise.

B. A three-wire microphone cable designed to minimize electrical interference.

C. The microphone component that converts changing air pressure into an electrical signal.

D. Noises produced by the shooting environment.

E. Replacing dialogue in real time by recording it synchronously with video playback.

F. A radio microphone.

G. "Live" audio recorded with the video.

H. A studio microphone support consisting of a rolling pedestal and a horizontal arm.

I. A two-wire microphone cable subject to electrical interference, used in amateur applications.

J. A device that converts sound waves into electrical modulations for recording.

K. A circuit that adjusts recording volume automatically.

L. A spatial pattern of microphone sensitivity, resembling a Valentine heart.

M. A microphone support consisting of a hand-held telescoping arm.

N. Sound as electronically recorded and reproduced.

O. The resistance of an electrical component to the passage of electricity.

P. A very small microphone clipped to the subject's clothing, close to the mouth.

Q. The directions in which a microphone is most sensitive to sounds.

R. Noises recorded and edited to match the video.

S. A device that balances the strengths of signals from two or more sources.

T. Without sound.

U. Previously recorded audio reproduced so that actors can synchronize their lip movements to it.

V. A specialized radio that picks up the output of wireless mikes.

W. A device for adjusting the relative strengths of different audio frequencies.

X. A written video or spoken audio identification of a program component.

Y. An area set up for recording real-time sound effects synchronously with video playback.

Chapter Quiz

Answer each question below. For True/False questions, circle the correct answer. For other questions, write the answer in the space provided.

1. The two big problems with audio recording are

_____

_____.

2. The five main types of sound recording are

_____

_____.

3. You fill gaps in the production track with

_____

_____.

4. The two main kinds of microphones are called

_____

_____.

T F 5. High impedance microphones deliver better quality sound.

6. In field recording, most independent microphones are suspended from

_____

_____.

7. Match the term to its definition by writing the definition letter in the space provided.

_____ Sound mixer   A. A control that balances the sound inputs from different
_____ Sound equalizer   microphones.
                            B. A control that balances the strengths of different sound
                               frequencies.

8. List two advantages of recording sound effects and other audio with a camcorder.

_____

_____

9. Three characteristics that distinguish good production audio are

_____

_____.

10. A room specially set up for recording sound effects is called a

_____

_____.

Activity 16-1

Mini-project: Recording Sound Effects

Description

This activity provides practice in recording audio effects in the field.

Objective

To obtain quality, noise-free recordings of sound effects.

Materials

- A camcorder kit.
- Headphones.
- Yellow pad and marker for slating (optional).

Preproduction

1. Select and write down two or three different sound effects to record. Identify their locations.
2. For each type of sound, select and write down at least three different versions of the effect. Here are some examples.

Car sounds

A. Door opening/closing
B. Engine starting, idling
C. Car driving away

Door sounds

A. Door opening, closing
B. Door slamming
C. Key inserted in lock, handle turned

Baseball (or softball) sounds

A. Ball hitting catcher's glove
B. Ball hit by bat
C. Slide into first base

Kitchen sounds

A. Water turned on, running, turned off
B. Oven door opened, closed
C. Spoon stirring bowl contents

Computer sounds

A. PC computer booting
B. Keyboard typing
C. Printer printing

Pet shop sounds

A. Dog barks
B. Fish tank bubbler
C. Parrot (or similar bird) "talking"

Production

Method

To optimize sound recordings for editing, you need to ensure that they are *realistic*, *isolated*, and *appropriate* for their environment.

- **Realistic** effects are easily recognizable and sound as they would if viewers heard them in real life. For example, if you record a car engine with the hood open and the microphone six inches away, the result may not sound like an engine at all. Listen to make sure the sound, as miked, is recognizable and realistic.
- **Isolated** effects are free from background noise that would make them hard to isolate and lay in during editing. Select quiet locations and monitor the background closely as you record and play back your sounds. Distant traffic, talking, or even birds often sound more obvious on tape.
- **Appropriate** sound effects match the environment in which they are supposedly heard. For example, if the visual is inside a car, then the engine effect should be recorded from inside; but if the visual is from the POV of someone watching the car leave, the microphone should be outside. Also, be aware of *presence*. An engine starting inside a metal storage building sounds much different from one started in the open air. Make sure your effects sound right for their screen environments.

Procedure

Here is the basic procedure for recording sound effects, using your camcorder as both microphone and recorder.

- Always listen through quality earphones that cover the ears and keep out exterior sound. For critical listening, you may need to press the ear pads tight against your ears. To facilitate recording, one person should operate the camera while another monitors audio quality.
- Slate every take to identify it. Usually, voice slates work well: "Car engine starting, take 1." If practical, frame a video image that helps identify the sound effect (like the grill and hood of a car). To record additional takes, stick two or more fingers into the frame for visual ID while saying the take number aloud. (If more than five takes are required, voice slate a second series and start over, for example, "Car engine starting, second series, take 1."
- Test various microphone positions by placing the camcorder in standby mode and trying different angles and distances. In general, the closer to the sound source you can get (without distorting the sound or risking injury), the "cleaner" the resulting track will be.
- As you record, monitor the sound carefully, and play it back after each series of takes.
- When recording more than one effect (say, of a car starting and leaving) keep the microphone in the same position for every effect.
- Keep a written log as you record. List each effect, leaving space at the left for take marks. For each take, place a "|" in the left margin. When you identify the best take, circle its mark, and shown in this example:

> | | | | | ① Car door opening, closing, from outside car. (Take 4 also good.)

The extra information is helpful in cataloguing and storing the effects for use in other programs.

- Finally, videotape a wide shot of the action—say, a person getting in a car, starting it, and driving away or a cook running water, beating a mixture in a bowl, and opening/closing the oven door to check something baking. This visual will provide footage to use in editing.

Postproduction

Repeat the following process for each of your two or three different types of sound effect.

1. Using your desktop editing system, start a project by importing the wide shot of the action for which the effects were recorded. Lay it down *twice*.
2. Import, name, and store the "circled take" (best take) of each sound effect recorded.
3. Remove the production audio from the second version of the wide shot and lay in the sound effects to synchronize with the visuals.

Discussion

After playing back the edited program, discuss the following questions.
1. How do the versions with production sound differ from those with laid-in sound effects?
2. Which versions sounds more realistic? Which sounds more dramatic? Why?
3. What seems to be missing from the sound effect versions? How would you add it?

Activity 16-2

The Literate Viewer: Identifying Sound Effects

Description

This activity will let you see how important sound effects are in dramatic programs.

Objective

To identify all the sound effects in about ten minutes of commercial television.

Materials

- TV set.
- VCR.
- Videotape.
- Yellow pad or notebook page.

Preproduction

Select a dramatic program to record (preferably one with considerable action, such as a police or hospital series).

Production

Tape record the program.

Postproduction

1. Select one "act" of the program (a story sequence from the end of one set of commercials to the start of the next).

2. Watch and listen to the sequence, pausing and repeating sections if necessary.

3. As you study the sequence, write down on the worksheet only the sounds you think are separately laid in sound effects, rather than sounds on the production track.

4. Also write on the worksheet the background noises that you think are background effects tracks, rather than the production track.

Discussion

1. Is it always possible to tell whether a sound is a separate sound effect? Which kinds of sounds seem more likely to be effects? Why?

2. Which background sounds do you think are separate effects tracks? Why do you think so?

3. Not counting separate stereo channels, a TV program may have a dozen or more sound effects and background tracks audible at once. Why do they use so many? Why not rely on the production track originally recorded with the action?

Sound effect (or bg track)	Sound effect (or bg track)	Sound effect (or bg track)	Sound effect (or bg track)
(sample) fire alarm			

Activity 16-3

Mini-project: Recording Sound with Built-in vs. External Microphones

Description

This activity helps you see the difference between the results obtained by the camcorder's built-in mike and a separate microphone.

Note: Some type of external microphone is required for this activity. It could be a directional mike on a boom, a handheld directional mike above or below the image frame, or a pair of lapel mikes and a production mixer.

Objective

To demonstrate the difficulties of recording with a built-in microphone and the advantages of using a separate microphone.

Materials

- Two subjects.
- External mike(s)—See note above.
- A camcorder kit.

Preproduction

Select a location where the two subjects can sit and hold a conversation—preferably outdoors, where some background noise is present.

Production

1. Using only the *camcorder's built-in microphone,* tape a two-shot of a short scene, such as the "Sample Script" nursery rhyme.
2. Tape the same scene again, over the shoulder of Subject B, framing Subject A.
3. Tape the scene a third time, framing Subject B over Subject A's shoulder.
4. Using whatever *external miking system* you have, repeat steps one through three, keeping the microphone in the same position, close to the subjects. Aim it at both subjects for the two-shot, then at A, and finally at B.

Postproduction

1. Using the shots made with the built-in microphone, edit the scene together, alternating pieces of all three setups.
2. Re-edit the scene in the same way, using the separately miked shots.

Discussion

1. What are the audio differences between tracks recorded with built-in vs. external microphones? Why do these differences occur?
2. Did the camcorder's automatic gain (recording volume) control affect sound levels? If so, how were the built-in mike's tracks affected? How were the external mike tracks affected?

Suggested Script

A: One, two...buckle my shoe.

B: Three, four...shut the door.

A: Five, six...pick up sticks.

B: Seven, eight...lay them straight.

A: Nine, ten...a great fat hen.

Directing for Content

Reading Review

Write a brief answer for each question below. If you are not sure of an answer, review the appropriate section of the chapter to find it.

1. Effective communication with an audience has three different aspects. What are they?

_____

2. Explain the difference, in directing, between shaping *communication* and *performance*.

_____

_____

3. Explain the difference between *subjective* and *objective* inserts.

_____

4. Explain the feelings imparted by wide angle and telephoto lens perspectives.

_____

_____

_____

5. How do different kinds of camera movement create different emotional effects?

_____

_____

_____

6. Why do performers need to be given the context of each shot?

_____

_____

7. Why does a teleprompter need to be very close to the camera lens?

_____

8. Explain why performer "marks" are often needed.

_____

9. What is the reason why performers (except spokespersons) should not look at the camera?

_____

10. Why do you think a low camera angle makes performers look more powerful and impressive?

_____

Vocabulary Review

Match each term to its definition by writing the letter of the correct definition in the space provided. (Not every one of these terms is defined in the *Technical Terms* section in the textbook, and some of the definitions do not apply to any term listed here.)

Terms

_____ 1. Throwaway question

_____ 2. Cover

_____ 3. Line reading

_____ 4. Wallpaper

_____ 5. Blocking

_____ 6. Subjective insert

_____ 7. Objective insert

_____ 8. Business

_____ 9. Cutaway

_____ 10. Pickup

Definitions

A. A short unit of action in a program, often (though not always) corresponding to a scene.

B. A shot other than, but related to, the main action.

C. A detail of the action presented from a character's point of view.

D. Places within the shot where the performer is to pause, stop, turn, etc.

E. A question asked not for the program, but to warm up the subject.

F. Footage designed to fill screen time while the audio delivers the important information.

G. A machine that displays text progressively as a performer reads it on-camera.

H. Referring to an off-level camera. The phrase "put dutch on a shot" is to purposely tilt the composition.

I. A detail of the action presented from a neutral point of view.

J. Additional angles of the main subject, or shots like inserts and cutaways, recorded to provide the material needed for smooth editing.

K. A performer's movement within a shot.

L. Scripted speech to be spoken by performers.

M. A shot obtained later to record action that was either missed or inadequately covered previously.

N. A vocal interpretation of a line that includes its speed, emphases, and intonations.

O. Activities performed during a shot, such as writing a letter or filling a vase with flowers.

Chapter Quiz

Answer each question below. For True/False or Multiple Choice questions, circle the correct answer. When more than one answer seems reasonable, choose the best one. For other questions, write the answer in the space provided.

1. A director must pay attention to these four things at once:

   _____ .

2. Communicating with an audience means delivering

   _____ .

3. List four ways to add emphasis to the information presented.

   _____

   _____

4. A "subjective" insert is taken from the point of view of

   _____ .

5. List five ways to enhance the emotional effect of a shot or sequence.

   _____

   _____

T   F   6. Generally, the bigger the image size, the more impact it has on the viewer.

7. List one tip for helping performers overcome self-consciousness.

   _____

T   F   8. Cue cards should always be prepared in advance, to be ready for use.

9. To help performers express complex emotions:
   A. Break them into parts and record each emotion in a different shot.
   B. Avoid complicated feelings in stories.

10. Low camera angles tend to make performers look

   _____ .

Activity 17-1

Mini-project: Directing for Emphasis

Description

This activity offers practice in emphasizing important information in a video sequence.

Objective

To use inserts, cutaways, and camera angles to direct audience attention.

Materials

- A camcorder kit.
- Lights, as needed.
- One or two performers, as needed.
- Props as specified in the treatments.

Preproduction

1. Select a sequence to direct and record from the treatment samples, or invent your own brief story.

2. Storyboard your sequence, with special attention to the setups intended to emphasize important story points.

3. Identify your shooting location, assemble props as necessary, and cast your actor. (If you use your own story, you may have more than one performer.)

Production

Stage and tape the sequence as you have storyboarded it. You need not follow the storyboard exactly, since you will get new ideas as you shoot.

Postproduction

Edit your sequence for presentation.

Discussion

1. Overall, how successful was your project? What was good about it and what might be improved?

2. How well did each camera setup do its job in delivering information and adding emphasis? (Consider image size, composition, and camera angle.)

3. If you could reassemble the production to shoot "pickup shots," (additional material recorded later to improve the sequence) what would you shoot a different way? Why? What new material might you shoot? Why?

Sequence Treatments

If you choose one of these suggestions, feel free to modify it any way you like. If you invent your own sequence, create one where proper *emphasis* is important to viewer understanding and enjoyment. Although pronouns are used for convenience, neither of these treatments is gender-specific.

The Assignment. When the instructor (off-screen voice only) asks for a homework or paper assignment, she looks in her backpack: No assignment! Calmly at first, then with increasing worry and then desperation, she digs in her backpack, extracts papers, books, lunch, wallet, (or whatever else you can think of). She sifts through all her stuff, looking for the assignment, drops things, picks them up, stuffs things back in her backpack, finally gives up. Then she looks at the empty seat next to her. Her assignment is already out and sitting on the chair arm (or seat).

The Necktie. Unaccustomed to tying a tie, he is determined to get it right. He ties the tie, but the knot is crooked; he pulls the tie off and re-ties it. Now the knot is okay but the back (thin) end hangs six inches lower than the front one. He starts over and does a good job, except that he pulls the knot too tight and practically chokes. Trying to loosen the knot, he ruins it. Giving up in disgust, he pitches the tie into a wastebasket, undoes his top shirt button, approves of the effect, and walks off.

Name_____

Date_____

Activity 17-2

Mini-project: Directing for Effect (Feeling)

Description

This activity offers practice in directing a video sequence to affect viewers' feelings.

Objective

To use camera angles, image size, shot duration, composition, lens focal length, and camera movement to arouse feelings in viewers.

Materials

- A camcorder kit.
- Lights, as needed.
- Several performers, as needed.
- Props as specified in the treatments.

Preproduction

1. Select a sequence to direct and record from the treatments supplied, or invent your own brief story.
2. Storyboard your sequence, with special attention to the setups intended to emphasize important story points.
3. Identify your shooting location, assemble props as necessary, and cast your actors.

Production

Stage and tape the sequence as you have storyboarded it. You need not follow the storyboard exactly, since you will get new ideas as you shoot.

Postproduction

Edit your sequence for presentation.

Discussion

1. Overall, how successful was your project? What was good about it and what might be improved?
2. How well did each camera setup do its job in adding to the emotional feeling?
3. If you could reassemble the production to shoot "pickup shots," (additional material recorded later to improve the sequence), what would you shoot a different way? Why? What new material might you shoot? Why?

Sequence Treatments

If you choose one of these suggestions, feel free to modify it any way you like. If you prefer to invent your own sequence, create one where a strong feeling is important to viewer enjoyment. Although pronouns are used for convenience, neither of these treatments is gender-specific.

Casting the Play (suspense). Bob and Bill are in competition for a role in the play. As the play director announces one part after another, the two grow more and more tense as neither is selected. After five or six parts have been announced, the lead role goes to one of them, while the other tries to conceal his disappointment. *Note:* This works best if you show the play director and the other cast members as they are chosen, cutting back to Bob or Bill, and cutting back and forth between them.

Catch (dynamic action). Helen and Mary are playing catch (with a Frisbee, a tennis ball, a basketball, or even something you don't normally play catch with). With each toss, one person tries to make it harder for the other to make the catch. (How they do this depends on the ball or other object they're playing with.) The contest grows hotter and the contestants work harder until finally one misses a catch and both of them stop, winded from their exertion. *Note:* Work with lens focal lengths, unusual angles, and shot lengths to convey a feeling of energy and effort.

Name _____

Date _____

Activity 17-3

Mini-project: Managing Emotional Transitions

Description

This activity provides practice in helping inexperienced actors portray complex emotions.

Objective

To record an emotional transition by breaking it into separate shots.

Materials

- A camcorder kit.
- Lights, as needed.
- Two performers.

Preproduction

Cast the roles of Instructor and Student, choosing someone with some acting ability as the Student. Rehearse the short script provided below.

Production

1. Tape the entire script as a two-shot.
2. Repeat the script as a closeup on the Instructor.
3. Repeat the script as a closeup on the Student.

In shots 1 and 3, make sure the Student's change from joy to horror is not an instant switch but a slow transition.

Postproduction

Edit the tape to create *three different versions*:

1. Present the entire script as an uninterrupted two-shot.

2. Edit the footage like this:

    A. **Two-shot**: *That sure is an amazing essay!*

    B. **Student CU**: (The **Student** reacts with joy and pride.) **Instructor's voice off-screen**: *Maybe the worst one I've ever read!* (The **Student**'s joy slowly turns to horror as we watch.)

3. Edit the footage like this:

    A. **Two-shot**: *That sure is an amazing essay!*

    B. **Student CU**: (The **Student** reacts with joy and pride.)

    C. **Instructor CU**: *Maybe the worst one I've ever read!* (Hold on the **Instructor** for a beat.)

    D. **Student CU**: Now the student looks horrified.

Discussion

1. Of the three versions, which one is most convincing? Why?

2. From the standpoint of effective drama, what is wrong with the two-shot only version?

3. How convincingly does the "student" change from joyful to horrified? Discuss the performance here.

4. Does the third version work best or not? Either way, discuss your reasons.

Note: It is always possible that the performer playing the student will make an expert, convincing onscreen transition from joy to horror. In that case, use this question instead of number 4, above.

5. On the one hand, showing the "student's" change is dramatically effective. On the other hand, inserting the "instructor's" CU adds punch to the criticism and allows the editor to control the timing of the "student's" reaction. Which is more important here? Why?

Script

Instructor: *That sure is an amazing essay!*

(The **Student** reacts with joy and pride.)

Instructor: *Maybe the worst one I've ever read!*

(The **Student**'s joy slowly turns into horror.)

Activity 17-4

Mini-project: Camera Angles and Actor Authority

Description

This activity shows how camera height can affect the apparent authority of a subject.

Objective

To demonstrate the effects of high and low camera angles.

Materials

- A camcorder kit.
- Lights, as needed.
- Two performers.

Preproduction

This is essentially a continuation of the preceding activity. If you have omitted that activity, do its preproduction. Use the script provided for the preceding activity.

Production

1. If you have not already done so, record the entire script as a two-shot.
2. Record the **Instructor's** closeup twice: once with the camera placed somewhat below eye level, and again with it placed above eye level. *Note:* These angles are only low or high enough so that they are obviously above or below subject eye level. Avoid extreme low/high angles.
3. Record the **Student's** closeup twice, exactly like the shots of the Instructor.

Note: Remember that "high" and "low" refer to the camera, not the subject.

Postproduction

Edit the sequence *twice*, once with the Instructor's low angle paired with the student's high angle; then again, with the instructor's high angle and the student's low angle. Each time, assemble the sequence following this shot list:

A. **Two-shot**: *That sure is an amazing essay!*

B. **Student CU**: (The **Student** reacts with joy and pride.)

C. **Instructor CU**: *Maybe the worst one I've ever read!* (Hold on the **Instructor** for a beat.)

D. **Student CU**: Now the student looks horrified.

Discussion

1. Is there a difference in feeling between the two versions? Why?
2. Which version seems to enhance the story line better? Why?

Bonus Activity

Camera height can be used to "grow" shorter actors (especially short male Hollywood movie stars). To see how this works, choose the tallest subject and the shortest subject available and tape them in conversation (repeat the previously used script, if you like). Treat the sequence as follows:

1. With the *tall* person playing one role, frame a neutral (eye-level) closeup and tape the entire scene.

2. With the *short* person playing the other role, frame a neutral (eye-level) closeup and tape the entire scene.

3. Edit the shots together, using alternating closeups.

Bonus Discussion

1. Is there any apparent difference in the heights of the two subjects? Why not?

2. If you also wanted a two-shot, how could you maintain this illusion?

3. If one or both subjects had to move a short distance, how could you maintain the illusion? (*Hint:* Either raise the bridge or lower the river!)

Directing for Form

Reading Review

Write a brief answer for each question below. If you are not sure of an answer, review the appropriate section of the chapter to find it.

1. In general, how do you obtain full *coverage* of the video material you are shooting?

_____

2. Explain the difference between *repetition* and *overlap* in obtaining coverage.

_____

_____

3. Briefly define the three major types of cutaways.

_____

_____

4. Briefly define *continuity of action*.

_____

5. Explain the fundamental idea behind screen direction.

_____

6. List three basic reasons for maintaining consistent screen direction.

_____

_____

_____

7. What are the special screen direction rules for scenes inside vehicles?

_____

8. List three ways to change screen direction during a sequence.

_____

_____

9. List three techniques for enhancing apparent depth in the image.

_____

_____

_____

10. Why is a zoom in or out not a true camera movement?

_____

11. Define the term "composite movement."

_____

12. List the parts of a moving composition.

_____

_____

Vocabulary Review

Match each term to its definition by writing the letter of the correct definition in the space provided. (Not every one of these terms is defined in the *Technical Terms* section in the textbook, and some of the definitions do not apply to any term listed here.)

Terms

_____ 1. Screen direction

_____ 2. Action line

_____ 3. Correcting

_____ 4. Establishing shot

_____ 5. Continuity

_____ 6. Zooming

_____ 7. Panning

_____ 8. Protection shot

_____ 9. Dollying

_____ 10. Cutaway

Definitions

A. Moving the entire camera horizontally.

B. Moving the entire camera up or down through a vertical arc.

C. The process of ensuring the edited program is a continuous presentation.

D. Moving the camera up or down in place on its central support.

E. Anticipating the need for shots that will cut together smoothly.

F. An imaginary line separating camera and subjects, to maintain screen direction.

G. Moving the camera by panning or tilting.

H. A wide shot near the start of a sequence, intended to orient viewers.

I. Pivoting the camera horizontally in place.

J. A shot other than, but related to, the main action.

K. Changing the lens' angle of view to fill the frame with a smaller or larger area.

L. Pivoting the camera vertically in place.

M. A shot taken to help fix potential problems with other shots.

N. Making small continuous framing adjustments to maintain a good composition.

O. The subjects' orientation with respect to the borders of the screen.

Chapter Quiz

Answer each question below. For True/False or Multiple Choice questions, circle the correct answer. When more than one answer seems reasonable, choose the best one. For other questions, write the answer in the space provided.

1. Everything that is videotaped is
   A. used in postproduction.
   B. raw material for future editing.
   C. shot out of order.

T   F   2. *Repetition* means shooting everything more than once.

3. *Overlapping action* means beginning a second shot by

_____.

4. The four principles that contribute to good *coverage* are

_____.

T   F   5. A *color shot* shows something different that relates closely to the action.

6. Match these types of cutaway shots with their definitions:

_____ Insert   A. An expressive shot of the environment.

_____ Reaction shot   B. A closeup of a detail.

_____ Color shot   C. The response of another performer to the action of one of them.

7. List the three main types of continuity.

_____

8. Name two of the shot types typically found in "classical" coverage.

_____

9. You keep the camera on one side of an invisible "action line" in order to

_____.

10. List two reasons for moving the camera during a shot.

_____

T   F   11. A well-executed camera move maintains the same speed throughout.

Activity 18-1

Mini-project: Styles of Coverage

Description

This activity lets you experiment with different approaches to directing and editing.

Objective

To create and evaluate different styles of camera coverage.

Materials

- Textbook.
- Two subjects.
- A camcorder kit.
- Lights, as needed.
- Props, as needed.

Preproduction

1. Study the *Styles of Coverage* section in Chapter 16 of the textbook. You will be creating an example of each coverage style discussed there.

2. Review the sample script below or else create one of your own.

3. Storyboard each version of your sequence. This step is important, both to design each coverage type in advance, and to ensure that you get enough camera angles to build all three types.

4. Select two performers and an appropriate location.

Production

Organize your shooting so that you do can re-use some angles in more than one coverage style:

1. Record the entire scene in each of the six basic camera angles discussed in the text, to create your "classical coverage."

2. Record the entire scene in all the alternative angles you plan for your "contemporary coverage" version.

3. Record any extra angles, inserts, cutaways, or other material intended for your "personal coverage" version.

Postproduction

Using and re-using your footage as needed, edit "classical," "contemporary," and "personal" versions of the sequence.

Discussion

1. Which type of coverage is easiest for viewers to follow? Why?

2. For most types of programs today, which type of coverage will generally be more useful? Why?

3. Which type of coverage is easiest to shoot and edit? Which is the hardest? In each case, why?

4. In directing actual programs, how close should your coverage stay to the examples presented here? Why?

Script

This script is a more "generic" version of the sequence used in the textbook. Though pronouns are used for convenience, the roles are not gender-specific. If you do not have stormy weather, stage the sequence in the shade. Then get shots of cloudy/stormy skies elsewhere and edit them in as needed. If necessary, change "rain" to "snow."

He and **She** are sitting outdoors, with textbooks and notebooks spread out between them.

He: Glad we had a chance to study for this test.

She: We'd better finish fast. It looks like rain.

He: (checking the sky) Maybe it'll hold off a while.

She: You willing to take that risk?

He: I guess not, with all these books and stuff.

She: I'll help you pick them up.

He and **She** start gathering up their work and stuffing it into backpacks.

Activity 18-2

Mini-project: Screen Direction

Description

This activity lets you use screen direction to convey two very different kinds of action.

Objective

To suggest, through screen direction, that the subjects in two similar sequences are doing quite different things.

Materials

- Two subjects.
- Folded papers for treasure hunt "clues."
- A camcorder kit.
- Lights, as needed.
- Props, as needed.

Preproduction

1. After reading the two treatments, prepare any props you will need.
2. Identify between six and ten locations nearby.
3. Storyboard the treatments, using the form on the next page. (Copy the form, if more sheets are needed.)

Production

Tape each of the stories, shooting just one shot in each location. In every shot, have the subject enter the empty frame at the start and leave the frame at the end. Show both subjects together only in the opening and closing shots of each sequence.

1. Shoot **The Great Treasure Hunt** with **Her** always moving in one screen direction and **Him** always moving in the opposite one.
2. Shoot **The Great Pursuit** with **Him** and **Her** always moving in the same screen direction.
3. Shoot **The Great Pursuit** a second time, this time thoroughly mixing up the screen directions for each subject. *Hint:* It will save time if you shoot the same setups for 2 and 3, simply changing the screen direction as needed.

Postproduction

Edit all three sequences. As you do so, experiment with the in-point and out-point of each shot. Sometimes you will want a subject to both enter the frame and then leave it. At other times, try starting with the subject already in frame and ending after the subject leaves. For still other shots, start with an empty frame, but cut away before the subject leaves the frame.

Discussion

1. How does screen direction help clarify each story?
2. What are the differences between the two versions of the pursuit story? How important is screen direction to telling the story? Why?

Sample Treatments

(Gender pronouns are only for convenience.)

The Great Treasure Hunt. He and She find a paper in a hiding place. (Throughout, improvise appropriate hiding spots for clues. If you like, you can make some of them humorous.) After reading the paper, he and she look at each other competitively, then race off in opposite screen directions. Alternating back and forth between Him and Her, we see each person find one clue after another, always moving in the screen direction they started with. In the final shot, they race into frame together from opposite sides (still following their screen directions) and find the "prize." It is so silly or worth so little that they toss it away and leave together.

The Great Pursuit. Sitting together, She grabs his soda (or any other prop) teasingly and runs off with it. He pursues her, leaving in the same screen direction. Through all the following shots, each subject passes through a location; then the other person passes through in pursuit. (*Hint:* change camera setups for the second person at each location, to avoid jump cuts.) Eventually, he catches up with her and grabs the soda back, but by this time the can is empty (or whatever ending you prefer).

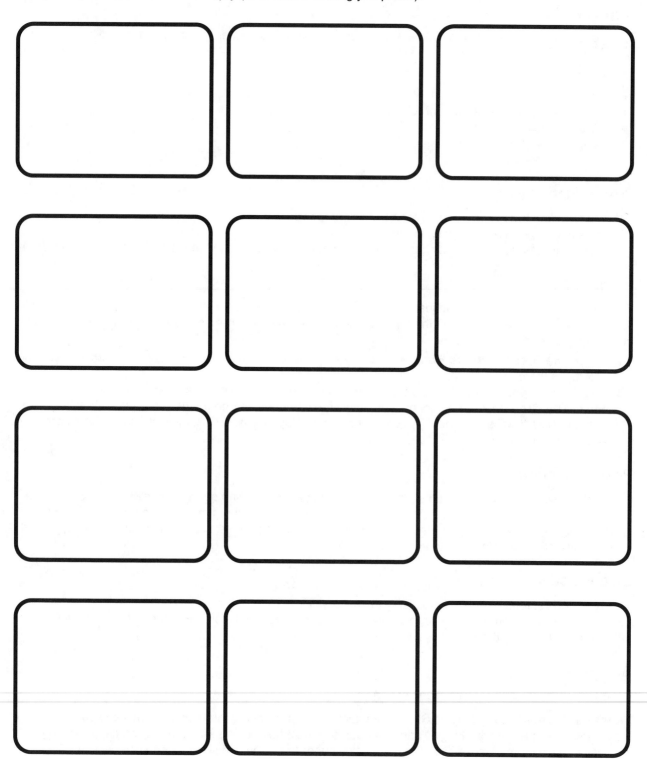

Activity 18-3

Mini-project: Changing Screen Direction

Description

This activity provides practice in changing a subject's screen direction, when necessary.

Objective

To master four different methods of smoothly changing screen direction.

Materials

- Two subjects.
- A camcorder kit.
- Lights, as needed.

Preproduction

1. Review the *Changing Screen Direction* section in Chapter 18 of the textbook.

2. Select a subject and a location that permits several different camera setups.

Production

Using a walking subject, shoot each type of direction change, as shown in the storyboards below.

Postproduction

Edit your four sequences as shown in the storyboards.

Discussion

1. How well do these four methods smooth over a direction switch? Do any work particularly well? Which ones and why?

2. Is maintaining screen direction always important? Is justifying a change in direction always necessary? In each case, why?

Storyboards

1. On-screen change.

2. Neutral direction.

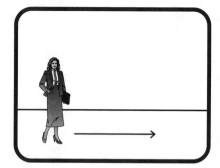  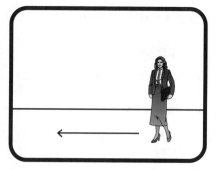

3. Cutaway buffer.

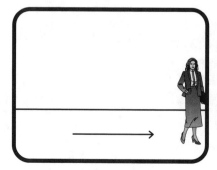 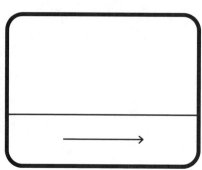 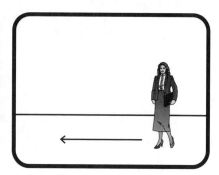

4. Empty frame..

Editing Operations

Reading Review

Write a brief answer for each question below. If you are not sure of an answer, review the appropriate section of the chapter to find it.

1. Define *the craft of editing* in its simplest form.

   _____

2. Explain the *subtractive* and *additive* approaches to editing.

   _____

   _____

3. Describe the differences between the two editing phases called *organizing* and *assembling*.

   _____

   _____

4. What do the four sets of numbers in a time code address refer to?

   _____

5. Explain the difference between *selecting* and *sequencing* shots.

   _____

6. Why does every shot have an *in-point* and an *out-point*?

   _____

7. Briefly explain what *conforming* means.

   _____

8. Explain what *rendering* a program means.

   _____

Vocabulary Review

Match each term to its definition by writing the letter of the correct definition in the space provided. (Not every one of these terms is defined in the *Technical Terms* section in the textbook, and some of the definitions do not apply to any term listed here.)

Terms

_____ 1. Slate

_____ 2. Timeline

_____ 3. Archiving

_____ 4. Additive editing

_____ 5. Enhancing

_____ 6. Organizing

_____ 7. Time code address

_____ 8. Composite

_____ 9. Subtractive editing

_____ 10. Filename

Definitions

A. The graphic representation of a program as a matrix of video, audio, and computer-generated elements.

B. Storing a finished program on disk or tape.

C. Choosing program components and copying them in order.

D. Creating a video program from production footage and other raw materials.

E. Adjusting program components for effect or to improve quality.

F. The unique code number assigned to each frame of video.

G. The identification of an editing element as it is stored in a computer.

H. The identification assigned a take before shooting it.

I. Creating multilevel images with compositions or superimpositions.

J. Creating a full-quality version of material previously edited in a lower-quality form.

K. Creating a program by starting with nothing and adding components.

L. A process in which elements of one image replace elements of another image to create a combination of both.

M. Cataloguing and filing program components.

N. Determining the correct order of program components.

O. Creating a program by removing material from the original footage and leaving the remainder.

Activity 21-3

Mini-project: Exploring DVEs

Description

This activity lets you experiment with the many DVE (digital video effects) included in your editing software.

Objective

To evaluate the character and quality of different DVEs and determine appropriate uses for them.

Materials

- A camcorder kit.
- Lights, as needed.
- Subjects, as needed.

Preproduction

Create and storyboard two different types of program material:

1. A narrative with a distinct time/space transition. For example, the subject finishes eating lunch and gathers up books and other possessions. The subject enters a classroom, sits down, and deploys a notebook and textbook.

2. A news program in which the newscaster introduces the sportscaster, who (after a couple of game reports) introduces the weather person, who starts reporting the weather.

Production

Tape the two mini-programs in preparation for editing.

Postproduction

1. Edit both programs, using only straight cuts.

2. Save four copies of the edited narrative program under different file names.

3. In the narrative program, make the transition from lunch to classroom in several different ways, using a different copy of the edited program for each one:

   - Fade-out/fade-in.
   - Dissolve.
   - Horizontal wipe (a vertical line sweeping across the screen). Experiment with hard-edged, soft-edged, and color-edged wipes, as well as with wipe directions (up and down, diagonal, etc.)
   - DVEs: depending on your software, select any fancy transitions you like, such as flips, fly-ins, page turns, and checkerboards.

4. Save four copies of the edited news program under different file names.

5. Make four different versions of the news program, using a different copy of the program for each one:

   - Select one DVE and use for both the transition from news to sports and from sports to weather.
   - Select two very different DVEs and use one for each transition.
   - Make the two transitions using slow horizontal wipes.
   - Make the two transitions using dissolves.

Discussion

1. In the lunch narrative, which transitions seem better—traditional ones or DVEs? Why?

2. How does a fade-out/fade-in differ from a dissolve or wipe in the way it communicates a lapse of time? Why?

3. In the news program, which transitions seem better—traditional ones or DVEs? Why?

4. Is there a difference in feeling between keeping the same DVE for both transitions or using different ones? Why or why not?

5. Where would you use a fade-out or fade-in during a news program? Why?

Activity 21-4

Mini-project: Exploring Split Edits

Description

This activity lets you explore different options in making split edits.

Objective

To learn the difference between "sound leading" and "picture leading" sound edits.

Materials

- A camcorder kit.
- Two subjects (plus one off-screen voice).
- Props, as needed.

Preproduction

1. Review the two kinds of "split edit" made when making a transition to a new scene:
   - *Sound leads picture:* The audio of the new scene begins over the ending of the old scene.
   - *Picture leads sound:* the audio of the old scene carries over into the start of the new scene.
2. Review the suggested script on page 212 (or write your own if you prefer).
3. Scout a student lunch area and a classroom.

Production

Shoot the script you have prepared. Try to get good-quality audio as well as video. Record the instructor's line as well. (Tip: have the performer stand about 15 feet from the camcorder so that the voice sounds a bit off-mike.)

Postproduction

Now edit the raw footage in three different ways:

1. *Straight edit*: keep sound and picture together in both scenes.
2. *Split edit, video leads*: Cut the video to the classroom *before* SHE says her last line, like this:

   HE slides into his seat and puts the backpack on the desk in front of him. Opening it, HE rummages around inside.

   SHE (off-screen): Boy is he going to be surprised when he opens that backpack!

   INSTRUCTOR (off-screen): All right, hand in your papers, please.

3. *Split edit, audio leads*: Lay the Instructor's off-screen line over the end of the first scene, as we see him walking away from the table, like this:

   SHE: Boy is he going to be surprised when he opens that backpack!

   INSTRUCTOR (off-screen): All right, hand in your papers, please.

   With an astonished look, HE pulls... (etc.).

Discussion

1. Do the three different transitions have different effects on the viewer? Explain your opinion.

2. Which version is most effective? Least effective? In each case, why?

3. There is no transition effect between these two scenes. With that in mind, what, how does either split edit affect the continuity of the program? Why?

Script

(As always, the script is not gender-specific. The pronouns are only for convenience.)

SCENE ONE: A LUNCH AREA

Two students are finishing lunch at a table with two backpacks on it. HE stands up in a hurry.

HE: Gotta run.

He grabs a backpack and indicates it.

HE: This paper's due at the start of class.

He turns and heads swiftly away.

SHE: Hey! That's the wrong backpack!

But He does not hear her and just keeps going.

SHE: Boy is he going to be surprised when he opens that backpack!

CUT TO:

SCENE TWO: A CLASSROOM

HE slides into his seat and puts the backpack on the desk in front of him. Opening it, HE rummages around inside.

INSTRUCTOR (offscreen): All right, hand in your papers, please.

With an astonished look, HE pulls a _____ out of the backpack and stares at it unbelievingly.

Mastering Digital Software

Reading Review

Write a brief answer for each question below. If you are not sure of an answer, review the appropriate section of the chapter to find it.

1. Explain why digital editing programs can be so complicated.

   _____

   _____

   _____

2. Why does a computer generally have to be prepared (optimized) for editing video?

   _____

3. What does a computer capture card do and when would you *not* need one?

   _____

   _____

4. What experience have you had with software tutorials and how well do they work for you? Why?

   _____

   _____

5. What is a storyboard? What is a timeline? What is the main difference between them?

   _____

   _____

6. Why would you want to change the scale of the timeline?

   _____

7. Why is the *Video Studio* workflow different from the one used generally in the book?

   _____

   _____

8. What is "rendering" and why is it necessary?

   _____

9. Why bother to trim clips before storing them?

   _____

10. In general, why is a logical filing system useful for large projects?

   _____

   _____

11. What is the advantage of editing one program sequence at a time?

_____

_____

12. Why preview a sequence after it is edited?

_____

13. Does previewing or rendering a sequence "lock it down"? Explain.

_____

_____

14. Explain what a plug-in is.

_____

15. How can a graphics management program be useful?

_____

16. How can you apply this chapter to the editing program you are using?

_____

_____

_____

Vocabulary Review

Match each term to its definition by writing the letter of the correct definition in the space provided. (Not every one of these terms is defined in the *Technical Terms* section in the textbook, and some of the definitions do not apply to any term listed here.)

Terms

_____ 1. Artifacts

_____ 2. Background track

_____ 3. Capture

_____ 4. Capture card

_____ 5. Clip

_____ 6. Drop frames

_____ 7. Effect

_____ 8. Filter

_____ 9. Frame-accurate

_____ 10. Swap file

_____ 11. Plug-in

_____ 12. Preview window

_____ 13. Render

_____ 14. Splitting

_____ 15. Storyboard

_____ 16. Thumbnail

_____ 17. Time code

_____ 18. Timeline

_____ 19. Toggle

_____ 20. Trim

Definitions

A. The method by which a user communicates with a computer.

B. To indicate the in- and out-points of a clip in order to specify its content and set its exact length.

C. A very small image representing a clip or shot.

D. To create a final, high-quality version of a finished program from the lower-quality "blueprint" of the project file.

E. A permanent hard drive area that holds frequently used data.

F. Short for digital video effect (DVE). Technically, any digital manipulation of the video, but commonly used to mean a scene-to-scene transition.

G. Small visual blemishes that mar the quality of an image.

H. An internal computer accessory card for capturing video.

I. A digital effect that changes the character—such as color—of the clip(s) it is applied to.

J. A window that displays what a piece of visual material looks (or will look) like.

K. In digital editing, a metaphor that represents the project as a succession of slides laid out in order.

L. In digital editing, a metaphor that represents a project as a stack of long, narrow strips, one strip for each program element.

M. A keystroke or keystroke combination that performs an action otherwise performed by a mouse.

N. An audio track of ambient sound, usually laid under the production audio track.

O. To fail to record one or more frames during capture, usually because of insufficient computer speed.

P. Edits made at exactly the single frame desired.

Q. In editing, separating the audio and video tracks of a clip so that each can be processed individually.

R. The address (location) of a clip, expressed in hours, minutes, seconds, and frames.

S. A series of closely related video shots, usually a few minutes long, or less.

T. Recording onto disk.

U. To switch back and forth between two states, such as off and on.

V. A program that has no interface of its own, but is operated by commands it has inserted into another program.

W. A unit of video, audio, graphics, or titles, as used in digital editing.

X. To import footage from a tape to a computer, digitizing it if the original is analog video.

Y. Any non-live-action visual element (but not commonly applied to titles).

Chapter Quiz

Answer each question below. For True/False and Multiple Choice questions, circle the correct answer. For other questions, write the answer in the space provided.

1. To process video editing more quickly, a computer needs _____.
   A. two monitors
   B. two hard drives
   C. two drive partitions

T F 2. You can edit video successfully on some laptop computers.

3. A "graphical user interface" is a _____.
   A. keyboard and mouse
   B. digital pen tablet
   C. system of commands in pictorial form

4. List four of the *Video Studio's* main screen parts.

   _____

   _____

5. Like the preview window in *Video Studio*, similar windows in most editing software programs can display at least three things:

   _____

   _____

T F 6. As you move through a program, the timeline moves from left to right.

7. When operating in storyboard mode, you set shot in- and out-points with _____.
   A. the timeline
   B. the navigation panel

T F 8. You can set the timeline scale so that it displays just one shot at a time.

9. Which of the following filenames would be the *least* specific in identifying a shot?
   A. 27 A 01.
   B. birthday 03 end trim.
   C. Marci closeup.

10. Define "plug-in."

    _____

T F 11. Previewing a sequence requires rendering a finished version of it.

12. A postproduction suite is a(n) _____.
    A. matched set of editing and similar programs
    B. edit bay where post production is carried out
    C. computer set up for video editing

13. List three functions of a graphics management database.

    _____

    _____

    _____

Name_____

Date_____

Activity 22-1

Learning Your System: Take a Tutorial

Note: The activities in this chapter are all intended to help you master the particular editing software that you are using. Depending on your program, you may need to adapt some of these activities to your situation.

Description

This activity lets you explore the interactive documentation available to you.

Objective

To evaluate tutorials, demonstrations, and similar learning aids.

Materials

- Disk or tape-based learning materials, supplied either by the software company or by third parties.

Background

This activity should tell you whether these tutorials are right for you—and perhaps whether you are right for tutorials in general.

Procedure

Take at least one full section of the materials supplied, following directions carefully.

Discussion

1. After taking the tutorial (or watching the demonstration) can you apply to your software the knowledge and skills covered? Explain.

2. Was the material too fast or too slow—too detailed or too sketchy? Explain.

3. Since tutorials and demonstrations run on a computer, just like the editing program, did you use your software while the tutorial was in progress? If so, how?

4. The textbook mentions several learning styles, including:

   - Reading materials before using the software.

   - Completing tutorials or other instructional aids.

   - Referring to reference guides and manuals when needed.

   - Experimenting with the software, discovering things as you go.

   Which learning style is closest to your own? Explain. What do you dislike about the other learning styles?

5. In your estimate, how long would it take to master the editing software completely? How long before you can start editing? Explain.

Activity 22-2

Learning Your System: Plot Your Workflow

Description

This activity compares your software to the textbook discussion.

Objective

To see how the textbook materials relate to your particular editing program.

Materials

- Textbook Chapter 22.
- Your editing program.
- The worksheet provided.

Background

As discussed in the text, most editing programs follow similar procedures, but they organize them differently and give them different names. Here, you can see how your program handles the editing workflow.

Procedure

The worksheet lists two editing workflows and the names of their steps. Analyze your software and enter its workflow steps, in order, in the third column. Notice that extra space is provided above and below to help you align your work with the others.

Discussion

1. How is your system similar to the other two? How is it different?

2. Why do you think editing programs have not standardized their procedures?

3. What things are most difficult for you to remember, find on the software interface, and use? Why?

4. All systems organize procedures in a straight line from configuring projects to archiving programs. Do you actually work that way? Could you organize a more realistic workflow? Explain.

Worksheet

Textbook	Video Studio	(your software) _____
1. Configuring digital projects		
2. Capturing materials	1. Capture	
3. Building the program	2. Edit	
4. Adding effects	3. Effect	
5. Designing the audio	4. Overlay	
6. Storing and distributing	5. Title	
	6. Audio	
	7. Share	

Activity 22-3

Learning Your System: Learn Your Interface

Description

This activity helps familiarize you with your editing screens and the features on them.

Objective

To learn the location of major screen working areas.

Materials

- Your editing program.
- Paper and pencil.

Background

The graphical user interfaces ("GUIs") created for editing programs have two major problems:

- They are too crowded.
- They are often not well identified.

To help overcome these problems, you will use paper and pencil as a learning aid to map the main screen areas of your software.

For example, as the textbook explains, Ulead's *MediaStudio* divides the work screen into seven major areas:

1. Step panel
2. Menu bar
3. Options panel
4. Preview window
5. Navigation panel
6. Library
7. Timeline

Each is always at a specific location on the screen (although contents may change slightly from one screen to another).

Procedure

1. Bring up the main editing screen of your software.
2. Using an 8 1/2 × 11 inch page as your "screen," use rectangles to draw the location of each major work area on your screen.
3. Label each work area.

Discussion

1. Compare the layout of your program to the many screen shots of *Video Studio* in the textbook. How are they similar/different? Why do you think your software looks different?
2. Many programs use two (or even more) preview screens, where *Video Studio* has only one. What are the pros and cons of each approach?
3. Many operations can be enabled by more than one method (such as clicking an icon, opening a right-button menu, using the keyboard). What are the pros and cons of this approach in learning the program? In using the program?

Authoring DVDs

Reading Review

Write a brief answer for each question below. If you are not sure of an answer, review the appropriate section of the chapter to find it.

1. Explain why DVD creation is called "authoring."

_____

_____

_____

2. In addition to video programs, name two of the other kinds of content that often appear on DVDs.

_____

3. Why is setting the TV standard an absolute must?

_____

_____

4. Name two types of content protection.

_____

_____

5. Explain the difference between the two types of content protection.

_____

_____

6. What is "digital linear tape" (DLT) used for?

_____

7. Explain "I-frames" and "GOP" in MPEG encoding.

_____

_____

8. Describe a "first play" clip.

_____

9. How can you guarantee that viewers will *not* skip the first play clip?

_____

10. What is a chapter thumbnail?

_____

_____

11. Explain what a playlist is.

_____

_____

12. Name two kinds of objects.

_____

13. List two mistakes that you look for in checking and previewing your design.

_____

_____

14. Name two of the three ways to archive a finished project.

_____

_____

Vocabulary Review

Match each term to its definition by writing the letter of the correct definition in the space provided. (Not every one of these terms is defined in the *Technical Terms* section in the textbook, and some of the definitions do not apply to any term listed here.)

Terms

_____ 1. TV standard

_____ 2. Regional encoding

_____ 3. MPEG

_____ 4. I-frames

_____ 5. DLT

_____ 6. Button

_____ 7. GOP

_____ 8. Authoring

_____ 9. Intra-frame compression

_____ 10. Chapter

_____ 11. Menus

_____ 12. Title

_____ 13. Scrub

_____ 14. Placeholder

_____ 15. Playlist

_____ 16. NTSC

_____ 17. Templates

_____ 18. Object

_____ 19. Disk image file

_____ 20. First play

Definitions

A. A recording format that records full information only for selected frames plus partial information for other frames.

B. A digital video recording system that saves storage space by condensing and recording the data for every frame.

C. Any program or menu.

D. The TV standard used in North America and Japan.

E. A computer disk format originally designed to hold movies, but now used for all-purpose data storage as well.

F. Mini-programs that guide users step-by-step through the menu creation process.

G. A system of encoding designed to prevent copying DVDs.

H. Pre-designed menus that are customized by substituting appropriate buttons and text.

I. To move through a program by dragging a handle along a track below a preview window.

J. A designated section of one program on a DVD.

K. A group of frames making up a complete unit of information in MPEG encoding.

L. "Intra-frames" that contain all the image data. (See MPEG).

M. A broadcast system such as NTSC.

N. An element on a DVD menu that, when activated, links to a program, a menu, or a function.

O. The process of designing and assembling DVDs.

P. On a menu template, an indicator of where an item should be placed to make it a button.

Q. An item on a menu that does not link to anything.

R. The most popular recording format for small camcorders.

S. Code on a DVD that permits it to be played in only one designated world region.

T. A DVD storage file format for recording on a special half-inch tape that will be the master for mass-producing DVDs.

U. Screens displaying buttons for navigating to other menus, programs, or functions.

V. A list for each button containing the names of one or more menus, programs, or functions activated by that button.

W. A computer file containing all the components of a DVD project.

X. The process of authoring a DVD.

Y. A menu or program that plays automatically when a viewer starts the DVD.

Chapter Quiz

Answer each question below. For True/False questions, circle the correct answer. For other questions, write the answer in the space provided.

T   F   1. The two main TV standards are 4×3 and 16×9.

2. Match the terms to their definitions:

_____ I-frame   A. A frame containing partial picture information.

_____ P or B frame B. A group of frames treated as an MPEG unit.

_____ GOP   C. A frame coded with all the picture information.

         D. The encoding system used to make DVDs.

3. Name three of the four types of information that are essential in configuring a new project.

_____

_____

4. What is a chapter?

_____

_____

T   F   5. Changing the thumbnail identifying a chapter will begin the chapter on the new thumbnail.

6. Name two common types of audio track.

_____

7. Name two common types of subtitle.

_____

8. Match the terms to their definitions:

_____ Object A. A connection between a button and a program, menu, or command.

       B. An inactive part of a menu.

_____ Button C. A part of a menu that links to other DVD components.

_____ Playlist D. A set of programs, menus, or commands activated by a button.

T   F   9. A placeholder is a frame around an object.

10. What does a "virtual DVD player" do?

_____

11. What are two forms in which you can archive a DVD project?

_____

Activity 23-1

Mini-project: Customizing a Template

Description

This activity allows you to convert a "canned" DVD menu into a menu specific to your program.

Objective

To master the process of customizing templates.

Materials

- Your authoring software.

- Appropriate templates.

- Two short programs. (*Tip:* in your editing software, you can make programs of just a few shots, to be used purely for practice.)

- An image (such as a .jpg photo) to use as a background.

Background

Most DVD authoring programs include templates that can be used instead of custom designs. At their simplest, they ask you to place your program material in preset buttons and use the rest of their design. However, you can also customize them by removing and replacing each element in turn, until you have created a semi-custom look.

Procedure

1. Select a DVD template that has positions for two or more program buttons.

2. Select the background and delete it. Import your still photo and place it as background.

3. Import your two programs into the appropriate work window. Drag each one into a button position. (Delete any additional buttons on the template.)

4. Selecting the buttons, drag them around the screen to suit the composition of your background.

5. Using the text function of your software, select each piece of text on your template and replace it with a typeface you have chosen for the program.

6. Use your program's testing procedures to verify that everything works properly.

Discussion

1. Many DVD templates look somewhat amateurish. Why is that?

2. In this activity, you changed the background, button positions, and text typeface. How effective were your changes in making the menu look "designed from scratch?" Explain.

3. Are there other elements on your page? How could you have customized them as well?

4. Would you consider using templates sometimes in your own work? Why or why not? If so, what projects would you use them for? Why?

Activity 23-2

Mini-project: Plotting a DVD

Description

This activity allows you to organize the links in a sample DVD.

Objective

To lay out a DVD with appropriate buttons on each screen.

Materials

- Component list.
- Organization chart.

Background

The component list below gives you the components of an imaginary DVD. The organizational chart provides hints about the menus on the DVD. The task is to assign appropriate buttons to each menu, including navigation buttons.

Procedure

1. Review the list of components that should be on the DVD.
2. Study the organizational chart provided.
3. Pencil in the DVD components on the appropriate menu(s).

Component list

- FRONT MATTER
  - FBI warning
  - Company logo clip

- PROGRAM
  - *Voyage to Tralfamador*
  - 5 program chapters
  - LANGUAGES
    - English
    - French

- SUBTITLES
  - French
  - Closed caption

- NAVIGATION AIDS
  - Setup
  - Play
  - Chapters
  - Main
  - Program

Flowchart

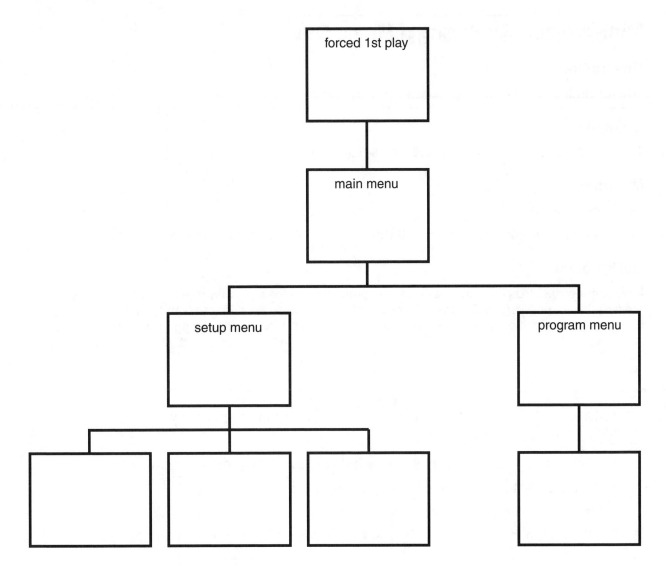

Activity 23-3

Mini-project: Analyzing a Movie DVD

Description

This activity lets you see how sophisticated DVD authoring can be.

Objective

To deconstruct a commercial DVD in order to reveal its design.

Materials

- A commercial DVD.

- A pad for notes and flowcharts. (See the Activity 23-2 for a typical DVD flowchart design.)

Background

Most commercial DVDs have complex menus—more complex than you may need in your own authoring projects. These professionally designed projects could have been created with high-level amateur/entry-level professional authoring software (probably including yours); so it is interesting to see what you can achieve.

Procedure

1. Choose a DVD. For simplicity, do not select a multi-disk DVD, a DVD with "Easter egg" (hidden) buttons, or a DVD that includes features intended for access by computer only.

2. Go through the DVD:

   - Note "forced first plays" that play automatically and cannot be exited until finished.

   - Note other first plays (usually previews), company logos, etc.

   - At the main menu, write down all the button choices.

   - Select one button and follow it to the next screen, note the buttons on this screen.

   - Follow each button to its lowest level menus, noting each one and its contents.

   - Return to the main menu and repeat the entire process with each of the remaining buttons.

   - Diagram the result.

Discussion

1. How many menu screens are there in the design? What is the average number of buttons on each screen?

2. How do navigation buttons ("home," "back," etc.) help provide random access to program material?

3. What extra features were used to enhance the DVD (such as moving buttons, moving backgrounds, sound effects, music loops, etc.). How much did they add to the overall effect? Why?

Analog Editing

Reading Review

Write a brief answer for each question below. If you are not sure of an answer, review the appropriate section of the chapter to find it.

1. Explain the terms "sequential" and "concurrent" with respect to analog video editing.

   _____

   _____

2. List the three essential parts of a simple editing setup.

   _____

3. Why is an editing monitor cabled to the source machine helpful?

   _____

4. How does analog video record electrical signals?

   _____

5. How does digital video record electrical signals.

   _____

6. Explain how a fixed erase head works.

   _____

7. Why is it useful to match jack and plug colors?

   _____

8. Studying Figure 24-7 in the textbook, determine the basic goal in cabling editing components together.

   _____

9. Explain the difference between a video insert and an audio dub.

   _____

10. Why does audio dub not work on stereo sound tracks?

   _____

   _____

Vocabulary Review

Match each term to its definition by writing the letter of the correct definition in the space provided. (Not every one of these terms is defined in the *Technical Terms* section in the textbook, and some of the definitions do not apply to any term listed here.)

Terms

_____ 1. Audio dubbing

_____ 2. A/B roll editing

_____ 3. Assembly tape

_____ 4. Jack

_____ 5. Assembly editing

_____ 6. Out-point

_____ 7. Preroll

_____ 8. Analog video

_____ 9. Linear editing

_____ 10. Random access

_____ 11. Flying head

_____ 12. Switcher

_____ 13. Runup

_____ 14. Generation loss

_____ 15. Source tape

_____ 16. Generation

_____ 17. Compositing

_____ 18. In-point

_____ 19. Insert editing

_____ 20. Digital video

Definitions

A. The difference between the point in time when the RECORD function is enabled and the first frame of recorded material.

B. Replacing a segment of previously recorded audio without disturbing the video.

C. Combining foreground elements from one image with the background of another.

D. A system that preserves audio/video information by sampling the electrical signal and recording the samples in binary (digital) form.

E. Starting the source deck playback prior to the in-point of the shot being transferred.

F. A record or erase head that operates in a continuous line on the tape passing across it.

G. The tape in the record VCR, on which the program is created.

H. A record or erase head that operates on the tape in short discrete diagonal segments.

I. An input and/or output port on a piece of equipment.

J. A transition or other effect created digitally, usually in a computer.

K. A system that preserves audio/video information by recording voltage fluctuations.

L. A level of tape copy.

M. An editing tool that selects and combines multiple sources of video and audio.

N. The decline in video and audio quality from one analog generation to the next.

O. The tape providing the raw material for editing (usually, but not always, an original camera tape).

P. The end of a cable to be inserted into a matching jack.

Q. A tiny electromagnet that encodes, decodes, or erases signals.

R. The first frame of a new piece of material.

S. Constructing a video program by transferring shots in sequence to a master tape.

T. Working with AV materials and finished program on videotape.

U. Replacing part of a previously edited video and/or audio segment with new material.

V. The last frame of a new piece of material.

W. Combining picture and track from two different program sources.

X. The ability to find and go to any part of a recording almost instantly.

Y. An informal name for assembly editing.

Chapter Quiz

Answer each question below. For True/False or Multiple Choice questions, circle the correct answer. When more than one answer seems reasonable, choose the best one. For other questions, write the answer in the space provided.

1. In linear editing, the process is both _____ and _____.
    A. analog
    B. concurrent
    C. sequential
    D. digital

T   F   2. Linear editing is analog and nonlinear editing is digital.

3. The two recorders (VCR or camcorder) used for linear editing are called the

_____.

4. A second monitor (TV) connected to the source deck is useful because

_____.

T   F   5. Unlike an analog recording, a digital recording is permanent.

T   F   6. Though used only in linear editing, a switcher is a digital appliance.

7. Flying erase heads are preferable to fixed erase heads because

_____.

8. The only editing port/cable combination that transmits data both in and out is called

_____.

9. Match the color coding of jacks and plugs to the signals they transmit.

_____ Red   A. S-video.
_____ White   B. Composite video.
_____ Yellow   C. Right channel audio.
_____ Black   D. Left channel/ mono audio.

10. Correctly sequence the steps in making an assembly edit by writing the numbers 1 through 5 on the appropriate blanks.

_____ Enable RECORD on assembly tape.
_____ Find new shot start frame.
_____ Position assembly tape.
_____ Set up source tape preroll.
_____ Roll source tape. Enable PLAY on the source deck.

11. The audio dub is typically used for

_____.

Name_____

Date_____

Activity 24-1

Mini-project: Cabling Editing Setups

Description

This activity offers practice in connecting editing components and troubleshooting problems.

Objective

To successfully create two different editing systems, finding and correcting cabling problems, as needed.

Materials

Materials will vary, depending on the equipment available. The following is a generic list.

- Source deck for original footage (either a camcorder with analog outputs or a VCR.
- Second source deck for B-roll footage.
- Processing unit (an A/B switcher or color processor).
- Record deck (VCR).
- Three TV monitors.
- Cables, four sets of each:
    - *RCA-plug audio cables:* white for left channel; red for right channel.
    - *Video cables:* either yellow RCA-plug composite or Y/C (SVHS).
    - *RF cables for monitors:* thick, round cables with bare copper wires in the centers of screw-on or push-on ends.

See Figure 24-6 in the textbook for examples of cable jacks.

Preproduction

1. Assemble your system components in a convenient place and clear a work area at least four feet wide. (It helps if you can walk behind your work table because most jacks are at the rear of components.)

2. Review Figures 24-2, 24-3, and 24-11 in the textbook.

Production

Following the diagrams on page 234, assemble as many of the setups as your equipment inventory allows. As you build your systems, keep these points in mind:

- The red, white, and yellow RCA coding on the equipment jacks always indicates the function of the jack. The color-coding on the cable plugs is only for convenience (any plug will work in any jack, regardless of color). However, matching plug and jack colors makes tracing cables much easier.

- RF cables are adequate for carrying picture and sound to TV monitors.

- If you use camcorders as source decks, the lack of RF jacks may prevent you from connecting your source decks to monitors. Instead, you can review your source footage as shown in Figure 24-2 in the textbook.

Postproduction

Disassemble and store components and cables, if appropriate.

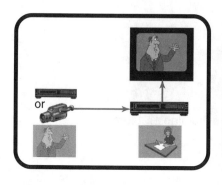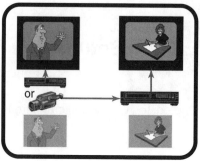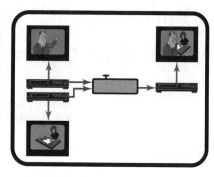

Discussion

1. If your cabling efforts required corrections in order to work, what were the problems?

2. What is the most logical way to trace a failure to get a signal to its intended destination? Why?

3. What are the advantages of having separate reference monitors for source decks as well as for the record deck? Why?

Activity 24-2

Mini-project: Calculating Runup

Description

This activity allows you to determine the lag time between the instant you enable RECORD and the first frame of material actually recorded.

Objective

To master the technique of making frame-accurate edits.

Materials

- A camcorder kit.
- Lights, as needed.
- Stopwatch or wristwatch with stopwatch function.
- Analog editing system (at least source deck, record deck, and TV monitor).

Preproduction

1. Carefully review the discussion of *Runup and Preroll* in the textbook.
2. Set up the camcorder to record a big closeup of the stopwatch. *Hint:* A telephoto lens setting will keep the camcorder far enough from the watch to allow shadowless lighting.

Production

1. Begin videotaping a closeup of the stopwatch set to 00:00:00:00, then start it running. Stop taping after the watch has counted perhaps 20 seconds.
2. Cue up the stopwatch footage in your source VCR at the head of the shot, before the watch is activated.
3. Cue up a second tape in your record VCR and place the deck in RECORD/PAUSE.
4. Play the stopwatch footage. At a selected point, such as 5 seconds exactly (00:00:05:00), enable RECORD on the assembly deck. Copy the running stopwatch for a few seconds.
5. Play back the shot on the assembly tape, noting the stopwatch time on the very first frame recorded (use slow motion or single frame advance, if possible). Suppose, for instance, that the reading is 00:00:06:50.
6. Subtract the start reading from the first frame reading:

     6:50 (first frame actually recorded)

   − 5:00 (frame at which RECORD was enabled)

     1:50 (runup time)

7. Repeat the process several times; average the runup times of all the shots, to arrive at a typical lag time for your equipment.

Postproduction

There is no postproduction.

Discussion

1. How close to one another were the various runup times?

2. What is "preroll" and how does it differ from runup time?

3. In editing matched-action shots, how do you use preroll and runup together?

Name_____

Date_____

Activity 24-3

Mini-project: Frame-accurate Edits

Description

This activity lets you practice "frame-accurate" edits, meaning that each new shot starts on exactly the frame you have chosen.

Objective

To consistently match action across edits, despite the uncertainties of "preroll" and "runup."

Materials

- A camcorder kit.
- Lights, as needed.
- Two subjects.
- An editing setup consisting of at least a source VCR and an assembly VCR, plus at least one TV monitor.

Preproduction

1. Configure your editing system: Cable source VCR to assembly VCR to monitor. (A reference monitor for the source VCR is convenient, but not essential.)
2. Review the Shot-to-shot Continuity section, in Chapter 20 of the textbook. Plan a simple action like the water pouring example in the textbook.

Production

Stage a simple action between your two subjects. You can use the textbook example, or invent a similar action of your own. Then record the following setups:

1. A master shot that includes both subjects.
2. A closeup of subject A.
3. A closeup of subject B.
4. A close shot (insert) of the action performed.

Make sure that every shot covers the entire action of your sequence.

Postproduction

Now, edit the sequence to match action, cutting from one shot to another during some sort of movement. (To help you make frame-accurate edits, review the section on *Runup and Preroll* in Chapter 24.)

Edit your raw footage in three different ways.

1. Start a new shot with the movement in the preceding shot out of the frame.
2. Start a new shot while the movement is paused.
3. Start a new shot during the actual movement, so that the incoming shot exactly continues the movement of the outgoing shot.

If your action is not extremely brief, at least two of your edits should involve movement in one of the ways noted above.

Discussion

1. How long does it take to learn frame-accurate editing with a linear system? What are some problems you encountered as you practiced?

2. Which of the three types of action-matching is easiest? Which is hardest? Why?

3. Which type of edit seems to do the best job of hiding the cut between two shots? Why?

Activity 24-4

Mini-project: Using Audio Dub

Description

This activity allows you to replace the live soundtrack on your program with alternative audio.

Objective

To "lay in" music or sound effects while leaving the video unmodified.

Materials

- A camcorder kit to tape footage for editing.
- An editing setup that includes an A/V mixer.
- A music track on tape, or on CD.

Preproduction

1. Configure your editing system: Cable source VCR to mixer to assembly VCR to monitor. Use the AUX input or similar if you need to plug a CD player into the mixer. An assembly VCR with AUDIO DUB capability is essential for this activity.

2. Plan a simple sequence, such as driving a car up and stopping, opening and closing the door, walking to a house door (or similar), opening the door, entering, and closing the door. (You may wish to storyboard this sequence to make sure you will cover everything.)

Production

You will record this sequence twice:

1. Tape the entire action, using several camera setups, each one appropriate to the particular action that it covers (such as closing the car door).

2. Tape exactly the same action again, this time placing the camcorder in the best position for recording each of the sound effects. (You may be able to do this as a single shot by hand-holding the camera close to the action, throughout the entire sequence.)

Postproduction

1. Edit your sequence in the usual way.

2. Make a working copy of your editing sequence. Cue it in the assembly VCR.

3. Using the audio dub function, lay the music track in under your edited sequence.

4. Cue a *second* working copy in your assembly VCR.

5. Cue your sound effects recording tape in the source VCR.

6. Using the AUDIO DUB feature on your assembly VCR (refer to *Using Audio Dub* in the textbook), lay down synchronized sound effects matching the video action. If repeated sounds like the footsteps prove too difficult, simply retain the ones from the production track.

Discussion

To hear the results of audio dub, remember to set your playback VCR to the MONO audio track.

1. How did the sound effects compare in quality to the sounds on the production track? What seemed to be missing?

2. How difficult was the process of laying down the effects? Why?

3. How well do you think the music track worked, as a substitute for production sound? Why?

Bonus activity

Just for fun, make a third working copy and cue it up as before. Only this time, plug a mike into the mixer and voice-dub the sound effects in real time. There are two ways to do it:

1. Make noises approximating the sounds on the production track, or

2. Say the *names* of the sounds in sync with the action. Laid in, the sound effects for the suggested sequence might sound like this:

SPEAKER: "Vroom. Screetch. Open. Close. Walk walk walk walk walk. Jingle. Open. Walk walk. Close."

Done carefully, the results can be quite funny.